W9-DDP-969

Bearded Protea
(Protea barbigera)

Una van der Spuy

South African Shrubs and Trees for the Garden

Hugh Keartland Publishers

HUGH KEARTLAND (PUBLISHERS) (PTY) LTD
Nicholson Street, Denver
Johannesburg

First edition 1971
Second edition 1975
This edition 1976
ISBN 0 949997 04 8

Filmset in 10 on 12 pt Century
Lithographed by Keartland Press (Pty) Ltd, Johannesburg

Contents

PART I. Introduction 7

The names of plants Habitat Frost Water Acid and alkaline soil
Measurements Acknowledgements Photographs
The meanings of some species names

PART II. Plant Selection Guide 17

Shrubs for the rock-garden
Shrubs and trees for coastal gardens
Shrubs and trees for a difficult climate
Shrubs for shady places
Shrubs and trees with fragrant flowers
Quick-growing shrubs and trees
Shrubs and trees with pleasing foliage
Shrubs and trees for growing in containers
Climbing Plants
Shrubs and trees which provide material for arrangements
Some trees for the small or large garden, patio, and for street and avenue planting
Trees and shrubs to provide colour through the seasons

PART III. Shrubs 61

PART IV. Trees 171

Index of Common Names 216

Index of Botanical Names 220

I

Part I
Introduction

The names of plants
Habitat
Frost
Water
Acid and alkaline soil
Measurements
Acknowledgements
Photographs
The meanings of some species names

◁ The Small Green Protea *(Protea scolymocephala)* is a charming species for the small garden.

Introduction

In many parts of Southern Africa beautiful shrubs and attractive trees are to be found growing wild— in forests, on plains, beside streams, on mountain slopes or along the coast. Some of them have a wide distribution inasmuch as they occur near the coast at low elevations and inland at high altitudes. One such beautiful tree is the Cape Chestnut *(Calodendrum capense)*, which can be found growing in forests from the Cape to central Africa. Other trees and shrubs are limited in their distribution, and can be found in the wild only in restricted areas. There is no reason, however, why gardeners in all parts of the country should not grow our indigenous shrubs and trees. In fact, it is astonishing that for so long we have furnished our gardens almost exclusively with exotic flora when such handsome plants grow in the veld around us. The reason for this neglect of our floral heritage is possibly due to the fact that for many years seeds and plants of indigenous flora have not been available from nurseries. This is no longer the case, however, for many nurseries now stock a wide range of South African plants, and there are even nurseries which specialize in growing only indigenous plants. Gardeners who wish to grow South African plants in their gardens or in pots and window-boxes, should join the Botanical Society of South Africa, the address of which is Kirstenbosch, Newlands, Cape. This organisation provides its members with a list of plants of which seed is available from the Society, and a journal which includes interesting articles on the growing of these plants.

In Australia and New Zealand many of our shrubs and trees are well known, and grown in private gardens and public parks; some of our proteas and ericas are cultivated in large numbers to produce flowers for florists in those countries, and for export to florists in other lands.

Names: Gardeners the world over find it difficult to remember the botanical names of plants, but for various reasons it is desirable to know them. In some countries many plants have common names which have been in general use for a long time, and one therefore tends to refer to these plants by their common names. Some African plants which have been cultivated abroad for years have been given common names in the countries where they have been grown, but often these common names differ from those used in South Africa. Furthermore, some of our wild plants have no common names, or else they have been given many different names; and sometimes, too, the same common name is used in different parts of the country for different plants. This leads to confusion, and, for the sake of clarity, therefore, I have arranged the plants in alphabetical order, based on their botanical names, with the common names following the botanical one.

As an understanding of the system used in naming plants makes it easier to remember the botanical names, some particulars about the system are given below.

Plants are grouped into different categories and named according to family, genus, species, variety and cultivar. The significance of the family name will be described last.

Genus: Plants which belong to a particular genus are closely related and one can often recognise this close relationship. Protea is a genus name and if you know what one protea looks like you can recognise others which belong to this genus. The genus name can be compared to the surname of a person, and where we use names such as Brown, McGregor or Cilliers, to identify people, we use genus names (plural genera) to identify plants. We speak of proteas, ericas or lebeckias. These are three genus names of three different genera of plants.

Species: The species (or specific) name of a plant may be compared to the first or Christian name of persons. If you wish to know to which McGregor or Cilliers someone is referring, you ask for his Christian or first name. In the same way, if you wish to know to which protea or erica someone is referring, you ask for the species (or specific) name.

Each genus may include one, or a large number of species. In the genus erica, for example, there are about six hundred different species. Very often the species name describes some characteristic of the plant. For example, one species of erica which has red flowers has been named *Erica coccinea,* because of the colour of its flowers, whilst another one, with sticky flowers, is named *Erica glutinosa.* Sometimes the species name is that of a person—generally an eminent botanist or plant collector, as in the case of *Erica sparrmannii* which is named after a Swede, Anders Sparrman, who arrived in South Africa in 1772, and who travelled many thousands of miles in many countries, botanising and collecting plants. Occasionally the species name is an indication of where the plant was first found or where it is common, as in *Tecomaria capensis.*

The genus name is always written first, starting with a capital letter, and the species name follows it, starting with a small letter. Where the genus name is mentioned again soon afterwards it is not necessary to write out the whole genus name again—the use of the initial letter is sufficient. For example, when writing about proteas, one may write about *Protea aristata* followed by some other protea, such as *P. compacta.* One does not write the whole word "protea" the next time it occurs.

Since it is usually easier to remember names if one knows their meanings, some of these are given at the end of this section.

Variety: All plants of a species resemble one another very closely indeed, although occasionally there may be slight differences, such as in the colour of the flowers. Where, however, there are other differences which are not sufficiently marked to merit another species name, the plant may be given a variety name. For example, we have *Erica coccinea* var. *pubescens* and *Erica coccinea* var. *melastoma.* They are both *Erica coccinea* because they so closely resemble the one originally named, but they differ very slightly from the original species named, and from each other.

Cultivar: Indicates a sport or hybrid cultivated for the garden. The cultivar name is always written in inverted commas and with initial capital letters, and it is placed after the species name. If, for example, another form of *Erica coccinea* were orginated, it would carry a name such as *Erica coccinea* 'Pink Beauty'.

Family: This defines a relationship among plants which concerns the botanist and plant collector rather than the gardener. The term "family" includes genera of related plants, and usually ends in "*aceae*", e.g. *Aizoaceae, Bignoniaceae, Ericaceae, Proteaceae.* The following shows more clearly the way in which these different terms indicating relationship are used. The family *Proteaceae* is divided into thirteen genera. Each of these genera is made up of different species. In the chart overleaf some species of only the genus mimetes are given.

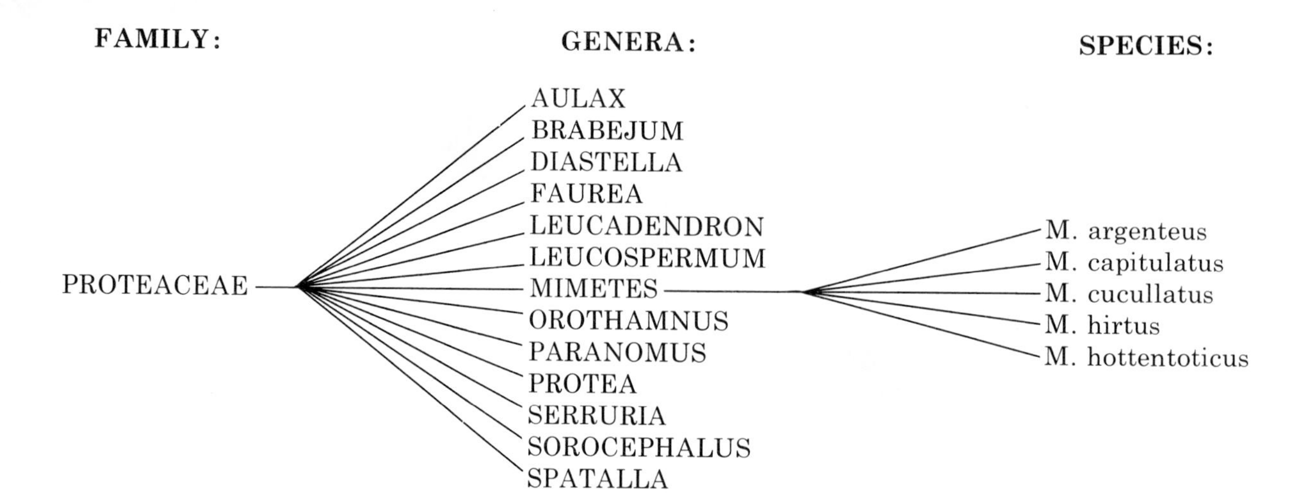

Changes of name: Unfortunately, the botanical names used for plants are subject to change. Such changes become necessary either because of the re-classification of plants or because it is discovered that a previous name was given. The rule followed by botanists is that the earlier name shall be accepted as the correct one and must supercede a botanical name given later, even although the later name may be more generally known amongst botanists and others. For example, one of the recent changes made in the names of proteas was the botanical name of the sugarbush. For many years this protea was known as *Protea mellifera,* the species name *mellifera* being a Latin word indicating that it secreted honey. This was an apt name, for the outstanding characteristic of this protea is the amount of sticky, honey-like nectar which it has. When, a few years ago, it was discovered that the name *Protea repens* had been given to this protea in the eighteenth century, the name *P. mellifera,* used for nearly 200 years, was changed back to *Protea repens,* despite the fact that the word *repens,* which means creeping, in no way describes the habit of growth of this plant.

Where in the text a second botanical name is given in brackets immediately after another, it is the name by which the plant was previously known.

Habitat: In the pages which follow, where individual shrubs and trees are described, the natural habitat of the plant is also given. This has been done to help gardeners provide the plants with conditions which suit them. If, for instance, a plant is from the south-western Cape, where the rains fall from autumn to spring, it should be watered regularly at this time of the year when it is grown in regions with a dry autumn and winter; and, where it is stated that a plant grows naturally in warm parts of Natal or the lowveld, gardeners will realise at once that such plants are unlikely to stand much frost.

Frost: Although on the whole winters are mild in most regions of Southern Africa, there are parts of the country where temperatures sometimes drop to below freezing at night, and there are a few areas where low night temperatures are the rule, night after night, for two to three months. As many of the shrubs and trees native to southern Africa grow naturally in areas where severe drops in temperature are unusual, these plants will need protection during winter when grown in gardens where low temperatures occur. Such protection is particularly necessary during the first three to four years. Once a shrub or tree is one or two metres (3-6 ft) tall it is less likely to be damaged by frost. This is because in winter temperatures at ground level may fall to ten degrees lower than they are at a metre above the ground.

Young shrubs and trees may be protected by having a "tent" or "wigwam" of hessian or straw erected around them for the winter months. This is an excellent way of providing protection and, once put up, it remains in place for the whole winter; however, some of the plants from the south-western Cape, particularly the proteas, like to have air circulating about them, and this permanent type of protection is not therefore suitable for them. The easiest way to shield them from frost is to invert an empty cardboard carton over them at night and to remove it during the day.

It should be remembered that very often the damage caused by frost is not due to the frost itself, but to the effect of bright and intense sun striking frosted leaves. This causes damage to the cells of the leaves, whereas if the plants are protected from the early morning sun in winter so that the frost is dissipated by the warmth of the air, such damage is unlikely to occur.

Tender plants—that is plants likely to be damaged by frost—will stand more chance of survival if grown where they are protected from cold by a wall, or by other plants. A wall facing the path of the winter-sun will absorb heat during the day and give off some warmth at night, so helping to keep plants near it from freezing.

Water: Although some plants die because of severe winters, more shrubs and trees succumb through receiving too little water than because of cold. Plants absorb their nutrients in solution, and a shortage of water therefore means that they may be starved even although they have been supplied with nutrients in the form of manure or fertilizer.

The commonest mistake gardeners make when watering is wetting only the surface of the soil. This moistening of the soil-surface gives the ground a pleasant appearance and may gladden the gardener's heart, but it does not help the plants, because their roots may be too far below the surface to be able to absorb the water. A thorough soaking of the ground twice a week takes less time and does far more good than a daily sprinkling of the surface of the soil. In order to enjoy one's garden it is advisable to work out a system of watering. Measure the amount of water the hosepipe delivers through the sprinkler, if a sprinkler is used, or through the end of the pipe, if the plants are watered individually without a sprinkler on the end of the hose, and try to ensure that each shrub and tree shall receive at least 14 litres (3 gallons) of water two or three times a week during the first three to four years of life. After this period many of them will have developed a deep root system and may need much less water. Trees whose habitat is a forest appear to do much better if they are watered by spray over the foliage rather than having water applied only to the soil.

Mulch: During periods of drought, and in regions where water is in short supply, the ground around shrubs and trees should be mulched. A 15 cm (6 inch) layer of straw or dry grass can be used for this purpose, but, if this is not available, cardboard, newspapers, plastic sheets or stones packed closely together on the ground around the plant will help to keep the moisture in the soil. Where a mulch has been put down, watering should be done by placing the end of the hosepipe under the mulch so that the water seeps into the ground beneath it.

Soil: The nature of the soil is another factor which should be considered when estimating how much water to give and how often. Sandy soils absorb water quickly and lose it quickly, whereas clay absorbs water slowly and retains it far longer than does sand; clay soil therefore should have more water applied at a time than is required for sandy soil, but it will not need to be watered as often.

Acid and Alkaline soil: A number of the plants described in this book occur in the south-western Cape, where the soil is acid, but many of them will grow in soils which are not acid, and which may, in fact, be slightly alkaline. Notable exceptions are members of the protea family and the ericas or heaths, most of which favour distinctly acid soil. If these are to be grown in areas where the soil is not acid, it is advisable to make large holes and to put acid compost and leaf mould and some peat into them, to ensure that these plants have the proper rooting medium. Where the water is brackish, even specially prepared soil may change in character because of the water, and the soil should be treated with regular dressings of aluminium sulphate or sulphur to keep it acid. A teaspoonful a month—sprinkled around the shrub which requires acid soil—is generally enough. Alkaline soils are more common in areas of low rainfall than in areas with a high rainfall.

The acidity/alkalinity state of the soil is measured in what is known as the pH. A pH of 7 is an indication that the soil is neutral, i.e. neither acid nor alkaline, whilst a pH reading below 7 indicates an acid soil, and one where the pH is higher than 7 shows an alkaline soil. Most plants prefer soil which is slightly acid—that is, with a pH of 6, and some, such as ericas, do best in soil which is much more acid, with a pH 4-5. Various firms and agricultural research stations do soil tests, for which they make a moderate charge. There are also soil-testing kits available for this purpose, and gardeners in areas where the soil is likely to be alkaline, and who wish to grow proteas and ericas should have the soil tested before planting them. Lime should not be applied to soil unless an analysis of the soil indicates that lime is required, and it should never be given to plants which like acid soil.

Measurements: Climatic conditions, which include proximity to the sea, elevation, rainfall and latitude play a part in the development of plants. So do other factors in the environment. For example, a tree which reaches a height of 30 m in a forest may grow to only half of this height in the open. The size to which plants grow, as given in the text, should therefore be regarded as approximate. The figures below serve as an aid in the conversion of metric to English measurements:

25.40 mm	=	1 inch
2.54 cm	=	1 inch
30.48 cm	=	1 foot
1 metre	=	3.28 feet
1 litre	=	0.22 gallons

Acknowledgements: I am much indebted to the staff of the Botanical Research Institute in Pretoria and to members of the staff of the Botanical Research Unit at Stellenbosch for checking the botanical names of plants described. I am also deeply indebted to the following:

Dr. J. Rourke of the Compton Herbarium, Kirstenbosch, who so willingly identified many plants and who also gave me much useful information on species with which I was not familiar; Mr. E. G. Oliver for help with the ericas; Mr. Ion Williams for reading through information and helping to identify the leucadendrons; Mr. and Mrs. F. C. Batchelor for allowing me to take photographs on their lovely estate; and Mr. and Mrs. G. Lipp for information on some Cape trees with which I was not familiar. I am grateful, also, to many persons who contributed information on their experience in growing indigenous shrubs and trees in their gardens in different parts of the country.

Photographs: In a book of this kind one would like to illustrate the nature of growth of every tree and shrub, and to show in addition, details of their flowers and leaves, but the inclusion of many pictures adds considerably to the cost of printing and one has therefore to limit the number of pictures included. The photographs are my own with the exception of the following: *Calodendrum capense,* for which I am indebted to Dr. F. Rousseau; *Encephalartos natalensis*, taken by Mrs. N. Gardiner; and *Ochna atropurpurea* by Mrs. C. Giddy; Prof. B. Rycroft for photographs of *Rhus lancea* and *Ziziphus mucronata*.

THE MEANINGS OF SOME SPECIES NAMES

acanthocarpa	with a thorny fruit
acaulis	stemless, or apparently so
acerosa	with needle-like leaves
aculeata	some part of the plant is prickly
acuminata	tapering
adianthifolia	with leaves like a maidenhair fern
adscendens	growing up or ascending
albida	white
alooides	like an aloe
amplexicaulis	stem-clasping
angustifolia	narrow-leaved
aphylla	without leaves
arborescens	like a tree
ardens	burning
argentea	silver
aristata	with a stiff bristle-like awn tapering to an apex
aurea	golden
auriculata	with ear-like appendages
baccans	becoming juicy and berry-like
barbigera	with a beard
bella	beautiful
blenna	mucus—referring to the stickiness of the flowers
brachiata	spreading, with branches in pairs
brachyandrus	with short anthers
brachypetala	with short petals
breviflorus	with short flowers
brevifolia	with short leaves
bubalina	buff-coloured
buxifolia	with leaves like the plant named box
caffra	from South Africa
calligerum	bearing beautiful flowers
calyptrata	capped
campanulatus	bell-shaped
candelabrum	like a candelabrum
cerinthoides	like cerinth (the honey-wort)

chamissonis . . . after von Chamisso—a German botanist and poet
chloroloma. . . . with a green fringe or edging
conocarpodendron . tree bearing cone-shaped fruit
cordifolium . . . with heart-shaped leaves
corifolia . . . . leathery leaves
cotinifolia . . . . having ovate-elliptic leaves, like cotinus
crenulata . . . . with wavy, indented margins
criniflorus . . . . with long, weak hairs
crinitum . . . . tufted
crithmifolia . . . with divided leaves
cryptopoda. . . . hidden stem (foot)
cucullatus . . . . hooded
cuneifolia . . . . with wedge-shaped or triangular-shaped leaves
curvirostris . . . with a curved beak
cuspidata . . . . sharply pointed at the apex
cynaroides . . . . like cynara (an artichoke flower)
cytisoides . . . . like the European broom

densifolia . . . . with crowded leaves
dichotoma . . . . branched
digitata. . . . . leaflets radiating like fingers from the hand
disticha. . . . . leaves or flowers in two ranks on opposite sides of stem and in same plane

erythrophyllum . . with crimson leaves
eugenea . . . . well-born
eximia . . . . . excellent
falcata . . . . . sickle-shaped
fascicularis. . . . in a bundle
fastigiata . . . . twiggy
floribunda . . . . with an abundance of flowers
formosa. . . . . beautiful
fragrans . . . . fragrant
frutescens . . . . growing in a shrubby fashion
fruticosa . . . . shrubby

gandogeri . . . . after Dr. Gandoger
genistoides. . . . like the European broom
glabrum . . . . smooth
glandulosa. . . . gland-bearing
globuligemma . . . with ball-like buds
glutinosus . . . . sticky
grandiceps . . . . large heads
grandis . . . . . large
gummifera . . . . bearing gum

haematoxylon . . . with wood the colour of blood
heteracantha . . . with different thorns
hirta . . . . . hairy
hispidum . . . . bristly
holosericea . . . silky-woolly
humeana . . . . low-growing
humilis . . . . . low-growing

ilicifolia . . . . with leaves like holly
imbricata . . . . overlapping as tiles on a roof
inerme . . . . . without spines
ingens . . . . . gigantic
inquinas . . . . fouled or stained
intrusa . . . . . inserted

lacticolor . . . . the colour of milk
lancea . . . . . lance-shaped
lanuginosa. . . . woolly, downy, cottony
latifolius . . . . broad-leaved
latissima . . . . very broad
leonitis . . . . . leonine
lepidocarpodendron . tree with scaly fruit
leucoptycodes . . . white folds
longicauda. . . . with long tails—referring to spurs on flowers
longiflora . . . . with long flowers
longifolia . . . . with long leaves
luteus . . . . . deep, golden-yellow

macrantha. . . . large-flowered
macrocarpa . . . with large fruits
macrocephala . . . with a large head
mauritanica . . . from mauritania
melanoleuca . . . black and white
melanthoides . . . very dark, black
microphyllum . . . with small leaves
microstigma . . . with small stigma
mucronata. . . . sharply pointed at the apex
multicava . . . . with many holes

muricata . . . . with short points like shell of murex

murex . . . . . like a murex shell
myrtifolia . . . . with leaves like a myrtle

nana . . . . . dwarf or small
natalensis . . . . from Natal
neriifolia . . . . leaves like oleander

obovata . . . . . blunt or rounded at the ends
oppositifolia . . . . with leaves opposite
oroboides . . . . with pea-shaped flowers like orobus

parilis . . . . . matched
paniculata . . . . with flowers in a panicle
parviflora . . . . small-flowered
patens . . . . . spreading
pavonia . . . . . like a peacock
pendula . . . . . hanging
pedunculata . . . with stalked flowers or leaves
peltatum . . . . peltate or shield-shaped
perfoliata . . . . leaves clasping stem
perfossa . . . . pierced through; hollow
petricola . . . . growing among rocks
peziza . . . . . cup-shaped
pinifolia . . . . with needle-like leaves,
pityphylla . . . . with leaves like a pine
physocarpa . . . with inflated fruit
procumbens . . . lying flat; trailing
prostratum . . . . growing prostrate—flat on the ground
puniceus . . . . crimson or phoenician purple

reflexus . . . . . turned back or reflexed
regia . . . . . royal
repens . . . . . creeping or low-growing

riparia . . . . . from the riverside
rotundifolius . . . having round leaves

rubropilosa . . . . red and hairy
rupicola . . . . rock-inhabiting plants

sarmentosus . . . with flexuous runners
scolymocephala . . edible, thistle-like head
sericea . . . . . covered with silky hairs
seriocephalum . . . with a silky head
sessiliflora . . . . having flowers without stalks
setosa . . . . . with bristles
sinuata . . . . . with wavy margins
spathaceus . . . . with bracts
spathulifolia . . . leaves like a spoon
speciosa . . . . handsome, showy,
speciosissimus . . . most handsome
spicata . . . . . bearing a spike
spinosa . . . . . with spines
squarrosa . . . . leaves jutting out at right-angles
subulatoides . . . looking like an awl

taxifolia . . . . with leaves like a yew
tetragona . . . . four-angled
tomentosa . . . . hairy
tortilis . . . . . twisted
transvenosus . . . cross-veining

uligonosum . . . growing in marshes
umbellata . . . . with umbels
uniflora . . . . . single-flowered
urticifolia . . . . with leaves like a nettle

vaginata . . . . sheathed
ventricosa . . . . inflated
verecunda . . . . right or genuine
vesicaria . . . . bladder-like, inflated
vestita . . . . . clothed
villosa . . . . . shaggy-haired
viminalis . . . . weeping
virgata . . . . . twiggy, long and slender
viscosa . . . . . sticky

xanthoconus . . . yellow cone

A fine arrangement of *Protea grandiceps*.

Leucospermum cordifolium, *L. conocarpodendron*, *L. tottum and Leucadendron* species.

Protea barbigera, *P. compacta*, *mauve Metalasia*, *Ericas*, *Phoenocoma* (pink everlasting), *Brunia* and *Watsonias*.

Proteas, *Leucospermums*, *Watsonias*, *Ericas* and leaves of the Silver Tree.

II

Part II
Plant Selection Guide

Shrubs for the rock garden
Shrubs and trees for coastal gardens
Shrubs and trees for a difficult climate
Shrubs for shady places
Shrubs and trees with fragrant flowers
Quick-growing shrubs and trees
Shrubs and trees with pleasing foliage
Climbing plants
Shrubs and trees for growing in containers
Shrubs and trees which provide material for arrangements
Some trees for the small or large garden, patio, and for street and avenue planting
Trees and shrubs to provide colour through the seasons

◁ The flowers of Shepherd's Delight *(Adenandra umbellata)* look as though they have been fashioned from the finest porcelain.

Plant Selection Guide

When reading a gardening book or on paging through a nurseryman's catalogue the gardener is often bewildered by the vast number of names—many of which are difficult to spell and to pronounce.

For this reason I have compiled lists of shrubs and trees which are more likely to succeed in a given situation than are others. These lists should not be considered exclusive but merely as a guide, for there are trees and shrubs other than those listed which will grow in an alien situation if given additional care.

Charts have also been compiled showing the season when different plants bloom; giving the names of those which have pleasing foliage or which provide material for arrangements, and those suitable for growing in containers.

Shrubs for the rock-garden .. 19
Shrubs and trees for coastal gardens .. 20
Shrubs and trees for a difficult climate .. 22
Shrubs for shady places .. 23
Shrubs and trees with fragrant flowers .. 24
Quick-growing shrubs and trees .. 25
Shrubs and trees with pleasing foliage .. 26
Climbing plants .. 27
Shrubs and trees for growing in containers .. 27
Shrubs and trees which provide material for arrangements .. 30
Some trees for the small or large garden, patio, and for street and avenue planting .. 31
Colour through the seasons .. 37
 Spring .. 37
 Summer .. 46
 Autumn .. 51
 Winter .. 53

SHRUBS FOR THE ROCK-GARDEN

A rock-garden of average size should be furnished with small plants—annuals, perennials, bulbs and succulents, but in the large rock-garden there is space for some shrubs, too. The ones listed below will contribute to the beauty of such a garden on a large estate or in a park, whilst some of them, such as the adenandras and the small ericas, will prove decorative in the small rock-garden, too.

Shrubs in a rock-garden should be sited with care so that they do not grow over the rocks and hide them from view. Where this happens it is sometimes possible to cut them back, but with some plants this may be impractical, and for this reason it is advisable always to measure the space allocated for each shrub, before planting.

The planning and construction of a rock-garden is not easy and its maintenance calls for constant vigilance. The soil should be prepared thoroughly before any planting is done and plants should be set out so that those which require an abundance of water are together, and those which will thrive with little water are planted in another section. If possible arrange to have some part of the rock-garden in shade so that shade-loving plants can be grown as well as the ones which like full sunshine.

SHRUBS

Adenandra species
Agathosma crenulata
Aspalathus species
Aster filifolius
Aulax species
Barleria obtusa
Berkheya species
Cadaba aphylla
Coleonema species
Encephalartos species
Erica (many species)
Erythrina acanthocarpa
Erythrina humeana
Erythrina zeyheri
Euryops species
Hermannia species
Hypericum revolutum
Indigofera species
Lachnaea species
Lebeckia species
Leucadendron (small species)
Leucospermum (small species)
Mackaya bella
Metalasia species
Mimetes cucullatus
Mimetes hirtus
Nymania capensis
Ochna atropurpurea
Oldenburgia arbuscula
Paranomus species
Phygelius capensis
Phylica pubescens
Plumbago auriculata
Podalyria sericea
Polygala virgata
Protea (small and low-growing species)
Rhigozum obovatum
Serruria species
Sutherlandia frutescens
Syncolostemon densiflorus
Thunbergia alata

SHRUBS AND TREES FOR COASTAL GARDENS

South Africa has an extensive coast line, with a climate which varies considerably from one part to another. The coastal strip of a good deal of South West Africa is cooled by sea mists and the sea current which sweeps up from the antipodes; in the south-western Cape cool winters and an abundance of rain during winter produce a splendid show of wild flowers during winter and spring; from the south-western Cape the climate gradually changes, becoming sub-tropical from East London into Natal and Zululand. Plants which thrive in one section of this coastal strip may grow fairly well in another, but some plants from the south-western Cape do not do well under hot humid conditions, whilst some from Natal do not flourish in the south-western Cape. Gardeners should therefore consult the text for further particulars about the plants listed here as being suitable for coastal gardens.

An important factor which influences the growth of plants along the coast is the nature of the soil. Very often it is sandy, and the nearer to the sea, the sandier it naturally becomes. In order to promote good growth gardeners near the sea should prepare the ground thoroughly before planting any shrubs and trees. Holes 60 cm (2ft) wide and deep should be made and filled with a mixture of compost, manure and peat, with some good soil. In coastal gardens wind may also prove a limiting factor and it is advisable to erect barriers to serve as a windbreak or to plant quick-growing trees and large shrubs for this purpose.

The growth of plants in coastal gardens also depends on how near to the sea the garden is situated. Although the plants listed will grow in coastal gardens only some of them are tolerant of salt-laden wind.

SHRUBS

Acokanthera oppositifolia
Acridocarpus natalitius
Adenandra species
Asclepias species
Aspalathus species
Aster filifolius
Barleria obtusa
Bauhinia galpinii
Berkheya species
Berzelia lanuginosa
Brunia species
Burchellia bubalina
Calpurnia species
Carissa species
Chrysanthemoides monilifera
Coleonema species
Crotalaria species
Cyclopia genistoides
Dombeya species
Dovyalis caffra
Duvernoia adhatodioides
Erica (some species)
Eriocephalus africanus
Euryops species
Gardenia amoena
Grewia occidentalis
Iboza riparia
Jasminum species
Leucadendron (some species)
Leucospermum (some species)
Lobostemon species
Melianthus species
Metalasia species
Mimetes cucullatus
Mimetes hirtus
Mundia spinosa
Ochna atropurpurea
Oldenburgia arbuscula
Paranomus species
Phygelius capensis
Plectranthus species
Plumbago auriculata
Podalyria species
Podranea species (climber)
Polygala species
Priestleya villosa
Protea compacta
Protea lepidocarpodendron
Protea neriifolia and others
Rhoicissus tomentosa (climber)
Senecio tamoides (climber)
Serruria species
Stoebe plumosa
Sutherlandia frutescens
Tecomaria capensis
Thunbergia alata (climber)

TREES

Alberta magna
Apodytes dimidiata
Brabeium stellatifolium
Brachylaena discolor
Buddleia salviifolia
Cassine crocea
Calodendrum capense
Cunonia capensis
Curtisia dentata
Cussonia spicata
Dais cotinifolia
Dodonaea viscosa
Dombeya rotundifolia
Ekebergia capensis
Erythrina caffra
Faurea macnaughtonii
Ficus species
Freylinia lanceolata
Gardenia species
Gonioma kamassi
Greyia species
Halleria lucida
Harpephyllum caffrum
Ilex mitis
Kiggelaria africana
Loxostylis alata
Millettia grandis
Mimusops caffra
Nuxia floribunda
Ochna pulchra
Olea africana
Olinia cymosa
Pittosporum viridiflorum
Podocarpus species
Rapanea melanophloeos
Rauvolfia caffra
Rhus chirindensis
Sparrmannia africana
Sideroxylon inerme
Syzgium gerrardii
Tarchonanthus camphoratus
Trichilia roka
Trimeria rotundifolia
Virgilia species
Ziziphus mucronata

SHRUBS AND TREES FOR A DIFFICULT CLIMATE

Although on the whole South Africa is blessed with a temperate climate there are parts of the country where low rainfall and a shortage of water for irrigation combined with arid air for most of the year, makes gardening difficult. It is interesting to note, however, that it is in some of these dry regions that masses of beautiful flowers appear naturally each spring to make vast carpets of colour in fields and on hillslopes. Most of our trees and many of our shrubs, however, occur nearer the sea or along mountain ranges where the rainfall is good. The ones listed below are more likely to endure harsh conditions than others and those marked with an asterisk are particularly hardy to drought. Some of them will also endure very cold winters. Unless there is an abundance of water for irrigation, gardeners living in dry areas should furnish their gardens mostly with plants which stand long periods without much water, but they should not limit themselves only to such drought-resistant subjects, because often, with a little extra attention, one can grow a variety of plants under unsuitable conditions. In areas of low rainfall it is advisable to put down a mulch around trees and shrubs to retain moisture in the soil. This may be a layer of straw about 10 cm thick or it may be of plastic sheeting, or of stones placed close together over the ground in a circle from near the stem of the plant to about a metre from it.

SHRUBS

*Asclepias species
Aspalathus species
Aster filifolius
Athanasia species
*Bauhinia galpinii
*Cadaba aphylla
Chrysanthemoides monilifera
Coleonema species
*Combretum species
*Dovyalis caffra
Erica blandfordia
Erica cerinthoides
Erica chamissonis
Erica chloroloma
Erica curvirostris
Erica grandiflora
Erica inflata
Erica mammosa
Erica peziza
Erica plukeneti
Erica taxifolia
*Erythrina acanthocarpa
*Erythrina zeyheri
*Lebeckia species
Leucadendron loranthifolium
Leucospermum conocarpodendron
Leucospermum muirii
Lobostemon species
*Metalasia species
*Nymania capensis
Paranomus species
*Plumbago auriculata
Podranea species
Polygala virgata
*Protea amplexicaulis
Protea caffra
Protea cynaroides and others
Rafnia ovata
*Rhigozum obovatum
*Senecio tamoides (climber)
*Sutherlandia frutescens
*Tecomaria capensis
Thunbergia alata (climber)

TREES

Acacia caffra
Acacia galpinii
*Acacia giraffae
Acacia haematoxylon
*Acacia karroo
Acacia tortilis subsp. heteracantha
Albizia harveyi
Brachylaena discolor
Brachystegia spiciformis
*Buddleia salviifolia
Cassia abbreviata subsp. bearana
Cussonia paniculata
Dodonaea viscosa
Dombeya rotundifolia
Erythrina caffra
*Euclea crispa
Greyia species
Kiggelaria africana
Kirkia acuminata
Lannea discolor
Olea africana
Pittosporum viridiflorum
*Rhus species
Salix species
Schotia afra
*Tamarix austro-africana
*Tarchonanthus camphoratus
*Ziziphus mucronata

SHRUBS FOR SHADY PLACES

One of the factors which limits the growth of plants in gardens throughout South Africa is a lack of shade. A large number of indigenous plants do much better if they are shaded for part of the day. Shade not only encourages their better development but it also often means less watering. There are some pretty shrubs which like shady places and there are others which do well when they are shaded for part of the day. It will be found that many of the shrubs which grow naturally in the south-western Cape will do better in hot inland gardens if they are planted so that they receive shade during the hottest hours of the day. This applies to many members of the protea family and to the ericas, too. Gardeners in areas which experience severe frost should remember also that early morning shade plays an important part in obviating frost damage. Plants marked with an asterisk do best in complete shade whilst the others prefer only light shade for part of the day when grown inland.

SHRUBS

Acridocarpus natalitius
Aulax species
Berzelia species
Burchellia bubalina
Carissa macrocarpa
Erica (many species)
Euryops acraeus
Gardenia species
Jasminum species
Leucadendron species
Leucospermum species
*Mackaya bella
Mimetes species
Melianthus species
*Phygelius capensis
*Plectranthus species
Protea (many species)
*Rhoicissus tomentosa (climber)
Rothmannia species
Senecio tamoides

SHRUBS AND TREES WITH FRAGRANT FLOWERS

The fact that one so often stops to smell a flower after having admired its beauty is proof of an inherent desire on the part of most of us to have plants with fragrance as well as beauty. Unfortunately many of our wild flowers have little scent. Apart from the freesias, a few species of crytanthus, crinum, babiana, gladioli, and *Tulbaghia fragrans*, our small plants are singularly lacking in perfume. Fortunately, however, we do have some shrubs and trees which bear flowers with a scent. Some of these will not do well in cold gardens whilst others will endure frost. Particulars of their culture are given in the text.

SHRUBS

Aspalathus species
Carissa macrocarpa
Clematis brachiata
Erica curvirostris
Erica peziza
Gardenia amoena
Jasminum species
Oncoba spinosa
Pavetta lanceolata
Pavetta revoluta
Podalyria calyptrata
Rothmannia capensis
Rothmannia globosa

TREES

Acacia galpinii
Acacia karroo
Acacia robusta
Acacia xanthophloea and others
Adansonia digitata
Afzelia cuanzensis
Burkea africana
Buddleia salviifolia
Calodendrum capense
Cassia abbreviata subsp. bearana
Ekebergia capensis
Faurea species
Freylinia lanceolata
Gardenia spatulifolia
Gardenia thunbergia
Gonioma kamassi
Holarrhena febrifuga
Ilex mitis
Khaya nyasica
Leucadendron argenteum
(Male Flower)
Parinari curatellifolia
Trichilia roka
Virgilia divaricata
Virgilia oroboides

QUICK-GROWING SHRUBS AND TREES

Some shrubs and trees are by nature slow-growing; others grow at a moderate rate and some are fast in their development. This is an innate characteristic of a genus or species. Climatic conditions and soil do, however, play an important part in promoting the better and faster growth of plants. A shrub or tree which grows naturally in a sub-tropical part of the country may develop well on the highveld or in the western Cape, but generally it will grow more slowly and with less exuberance than in its native habitat. Similarly, trees and shrubs which like cool growing conditions and some frost may not prove as fast in growth when planted in a hot climate. Gardeners should be prepared to experiment for this brings rewarding results. Fine proteas are grown within sixty kilometres of Durban and also in the Karoo, and on the highveld — regions which have a climate very different from that of their natural home.

Two factors which tend to retard growth are poor soil and a shortage of water. To promote quick growth dig holes about 60 cm deep and across and put in plenty of manure, compost and peat with some good soil. Manure and compost promote the multiplication of beneficial bacteria and peat absorbs and retains water around the roots of the young plants. Where possible water should be supplied to young plants at regular intervals and in such quantities that the water will seep down to their roots. Sprinkling the surface of the soil will not encourage shrubs and trees to grow. They should receive twelve to sixteen litres twice a week. This is particularly necessary during periods when they are actively growing—i.e. when they are in leaf and producing flowers. Unless one is prepared to provide plants with good soil and enough water to promote growth even the quick-growing ones may prove slow. Once they are mature many of them will require far less water as their roots will have reached down to draw on the humidity below the surface of the soil.

SHRUBS

Asclepias species
Aster filifolius
Athanasia species
Barleria obtusa
Bauhinia galpinii
Berkheya species
Chrysanthemoides monilifera
Coleonema species
Crotalaria capensis
Dombeya species
Erica (some species)
Jasmine species
Lebeckia species
Leucospermum (some species)
Metalasia species
Phygelius capensis
Plectranthus species
Plumbago auriculata
Podalyria species
Podranea species (climber)
Polygala species
Protea (many species)
Senecio tamoides (climber)
Sutherlandia frutescens
Tecomaria capensis
Thunbergia alata (climber)

TREES

Acacia albida
Acacia galpinii
Acacia sieberiana var. woodii
Albizia adianthifolia
Buddleia salviifolia
Cassine crocea
Cunonia capensis
Curtisia dentata
Erythrina species
Freylinia lanceolata
Halleria lucida
Harpephyllum caffrum
Nuxia floribunda
Olinia cymosa
Sparrmannia africana
Spathodea campanulata
Trimeria rotundifolia
Virgilia species

SHRUBS AND TREES WITH PLEASING FOLIAGE

Many gardeners are guided in their selection of plants by the flowers which they have. This is a mistake, for many shrubs and trees bear flowers for a comparatively short time—three to five weeks. If their foliage is unattractive they should be planted in such a position that they will not be obvious when not in flower. The owners of small gardens should be particularly careful to evaluate the foliage of shrubs and trees before choosing any for their gardens. Foliage may be attractive because of its form, its arrangement on the plant, its texture or its colour. In some cases the foliage is attractive only at certain seasons, as when it turns colour in spring or autumn, whilst in other cases it may be decorative throughout the year.

More detailed information about the plants listed below is included in the text.

SHRUBS

Acridocarpus natalitius
Burchellia bubalina
Carissa macrocarpa
Crotalaria capensis
Dombeya burgessiae
Encephalartos species
Euryops pectinatus
Euryops hybrid (unnamed)
Gardenia species
Jasminum species
Leucadendron (some species)
Leucospermum (several species)
Melianthus species
Metalasia rhoderoides
Ochna atropurpurea
Oldenburgia arbuscula
Oncoba spinosa
Phylica pubescens
Plumbago auriculata
Podalyria sericea
Protea (some species)
Rhoicissus tomentosa (climber)
Rothmannia species
Senecio tamoides (climber)
Tecomaria capensis

TREES

Acacia (several species)
Adansonia digitata
Afzelia cuanzensis
Alberta magna
Albizia adianthifolia
Brachylaena discolor
Brachystegia spiciformis
Calodendrum capense
Cassine crocea
Celtis africana
Cunonia capensis
Curtisia dentata
Cussonia species
Diospyros whytei
Fagara davyi
Faurea macnaughtonii
Gardenia spatulifolia
Gardenia thunbergia
Gonioma kamassi
Kirkia acuminata
Kirkia wilmsii
Leucadendron argenteum
Loxostylis alata
Millettia grandis
Nuxia floribunda
Ochna pulchra
Olea capensis
Olinia cymosa
Peltophorum africanum
Podocarpus henkelii
Rapanea melanophloes
Rauvolfia caffra
Rhus chirindensis
Schotia brachypetala
Sparrmannia africana
Spathodea campanulata
Trema orientalis
Trichilia roka
Trimeria rotundifolia
Vepris undulata
Virgilia species
Ziziphus mucronata

CLIMBING PLANTS

Although there are many decorative plants indigenous to Southern Africa, few of these are climbers. The list below includes plants which, strictly speaking, are not true climbers, but they are plants which need support to hold them erect.

NAME OF PLANT	REMARKS
Aloe ciliaris	A scandent aloe with tomato-red flowers which scrambles over any support available.
Aloe tenuior (FENCE ALOE)	Is commonly called a Fence Aloe because the long, rangy stem needs support.
Asparagus setaceus (ASPARAGUS FERN)	Grows to about 2 metres, with a slender stem and delicate fern-like leaves.
Clematis brachiata (TRAVELLER'S JOY)	This is a scrambling shrub which drapes itself over any support and bears sweetly-scented flowers.
Gloriosa species (FLAME LILY)	Grows from 1-1.5 metres and bears very handsome and unusual flowers of yellow shaded to crimson, in summer. Tendrils at the ends of leaves cling to any support available.
Jasminum multipartitum (JASMINE)	This jasmine is a handsome plant with dark, glossy leaves and fragrant white flowers. It makes a fine hedge if trained along a fence, but can be trained up a wall, too.
Podranea brycei and P. ricasoliana (ZIMBABWE CREEPER AND PORT ST. JOHN'S CREEPER)	These are vigorous plants which grow very quickly and cover a pergola or wall within two years. The pink flowers make a fine show in late summer and early autumn.
Rhoicissus tomentosa (WILD GRAPE, MONKEY ROPE)	Does not have showy flowers but is an evergreen with large, glossy leaves which are effective throughout the year.
Senecio tamoides (CANARY CREEPER)	This is a very quick-growing climber with attractive foliage and gay, yellow flowers in autumn. It stands considerable drought.
Thunbergia alata (BLACK-EYED SUSAN)	Black-eyed Susan is a quick-growing climber which produces masses of orange flowers on and off from spring to autumn. Hybrids have flowers of other colours as well.

SHRUBS AND TREES FOR GROWING IN CONTAINERS

Plants grown in containers have been popular for generations. They are of decorative value inside the house and they can also add to the beauty of the entire garden. In Europe it has been the practice for many years to cultivate frost-tender plants in a glass-house in winter and to bring the plants out to embellish the garden in spring and summer. Very often the containers used are large, measuring 60 cm and more across and in depth. By growing plants in containers one can be sure of having plants at their best, for, when one plant is past its prime it can be removed to an out-of-the-way part of the garden and another one in a container may be brought forward to take its place. Such plants, grown in containers, may be used to highlight the verandah or stoep, on a sunny or shady terrace, in a patio, at the top of a flight of steps or beside a pool.

Gardeners who live in areas where soil conditions make gardening difficult can cultivate many plants in containers of different sizes for different parts of the garden. Given a large enough container and the right soil mixture, one can grow a wide variety of shrubs, trees and climbers.

The larger the container the better, naturally, will be the growth of the plants, but whatever the size of the container it should be remembered that the plants grown in them will need regular applications of manure, or compost and fertilizer, to promote growth, and that they will also need to be watered far more often than plants growing in the garden. It is better to over-water rather than under-water, but, when giving fertilizer to plants in containers rather give a little at frequent intervals than a large amount every three or four months.

The containers chosen should suit the nature of growth of the plant. A small plant such as adenandra will look most attractive in an ordinary flower pot, but trees and shrubs generally grow better and look better when planted in large containers such as tubs or troughs which measure more than 60 cm in height and width.

Large container—with diameter more than 45 cm.
Medium container—with diameter of 30-45 cm.
Small container—with diameter of 20-30 cm.

NAME OF PLANT	SUN OR SHADE	LARGE	MEDIUM	SMALL	REMARKS
Adenandra species	Sun or part shade		•	•	Low-growing with pretty flowers from late winter to early spring.
Aspalathus spinosa	Sun	•	•	•	Yellow flowers in spring.
Aster filifolius	Sun or part shade	•	•		Stands dry conditions. Flowers in spring.
Barleria obtusa	Part shade	•	•		Exuberant-growing plant with mauve flowers in autumn.
Burchellia bubalina	Shade	•	•		Attractive foliage; flowers in spring.
Coleonema album and C. pulchrum	Sun	•	•		Can be trimmed into shapes. Flowers from late winter to mid-spring.
Cussonia species	Sun or part shade	•	•		Quick-growing, with most attractive leaves. A fine plant for the patio or terrace.
Dombeya tiliacea	Sun	•	•		Large leaves more decorative than its flowers.
Erica	Sun or part shade	•	•	•	Most require distinctly acid soil. Charming plants for a patio.
Euryops acraeus	Sun or part shade		•	•	Stands frost but not dryness. Yellow flowers.
Gardenia species	Part shade	•			Good foliage and scented flowers.
Jasminum species	Part shade	•			Good foliage and scented flowers.
Leucadendron argenteum	Sun or part shade	•			Silver leaves throughout the year. Requires acid soil.

NAME OF PLANT	SUN OR SHADE	LARGE	MEDIUM	SMALL	REMARKS
Leucadendron floridum	Sun or part shade	•			Pretty foliage and flowers.
Leucospermum cordifolium	Sun or part shade	•			Long flowering period from winter to spring.
Leucospermum tottum	Sun or part shade	•			Flowers of dusty pink in late spring. Requires acid soil.
Mackaya bella	Shade	•			Flowers of soft mauve in spring.
Plumbago auriculata	Sun or part shade	•	•		Quick-growing in poor soil. Stands drought. Blue flowers in summer.
Podocarpus henkelii	Sun or part shade	•	•		Attractive foliage throughout the year.
Podalyria sericea	Sun		•	•	Decorative silver leaves and mauve flowers.
Proteas. Little is known of their performance as container plants, but the following species should be tried. Protea aristata, P. cynaroides, P. grandiceps, P. longifolia, P. minor, P. pityphylla, P. scolymocephala, P. speciosa, P. sulphurea.	Part shade	•			Acid soil and an abundance of water during autumn, winter and spring are necessary for proteas.
Rhoicissus tomentosa	Part shade	•			Handsome foliage. Plant needs support.
Senecio tamoides	Sun	•	•		Climber needs support. Yellow flowers in autumn.
Serruria species	Part shade	•	•		Produce their delightful and unusual flowers from autumn to spring.
Sparrmannia africana	Sun or part shade	•	•		Large, velvety leaves which are decorative
Sutherlandia frutescens	Sun	•	•		Good foliage and red flowers followed by decorative pods.
Tecomaria capensis	Sun	•			Orange or yellow flowers in late summer and early autumn.
Thunbergia alata	Sun	•	•		Climber or cascading plant with orange flowers from spring to autumn.

SHRUBS AND TREES WHICH PROVIDE MATERIAL FOR ARRANGEMENTS

NAME OF PLANT	MATERIAL
Adenandra species (CHINA FLOWER, SHEPHERD'S DELIGHT)	Palest pink, rose or white flowers for small arrangements. (Winter and early spring).
Asclepias species (WILD COTTON)	Green inflated seed-pods. Burn ends of stems before arranging. (Summer)
Aulax species (AULAX)	Yellow male flowers of aulax for arrangements of medium size. (Late winter and early spring)
Berzelia lanuginosa (KOLKOL)	Rounded balls of ivory colour. (Late winter and early spring)
Brunia nodiflora (STOMPIE)	Ivory heads like little pincushions. (Late winter)
Coleonema album and **C. pulchrum** (CONFETTI BUSH)	Very small white and pink flowers. (Early spring)
Dombeya rotundifolia (WILD PEAR)	Blossom in heads like that of the pear. (Winter)
Erica or Heath (MANY SPECIES: SEE TEXT)	Flowers large or small of different shapes and colours. (They flower at different seasons; see text)
Eriocephalus africanus (WILD ROSEMARY, KAPOKBOSSIE)	Fluffy grey/white heads. (Spring)
Greyia sutherlandii (MOUNTAIN BOTTLEBRUSH)	Gleaming, crimson to scarlet flowers. (Late winter and early spring)
Holmskioldia tettensis (MAUVE CHINESE HAT FLOWER)	Long stems with small mauve flowers. (Summer)
Iboza riparia (IBOZA)	Minute flowers in misty mauve masses. (Winter)
Leucadendron species (SILVER TREE AND OTHERS)	Many species are decorative. The colours are silver, yellow, rose and green. (Late winter and early spring)
Leucospermum species (PINCUSHION AND OTHERS)	Several species are decorative but the best is undoubtedly *L. cordifolium*. The colours are salmon-pink, rose, apricot and coral-red. (Late winter and spring)
Mimetes species (MIMETES, SOLDAAT)	Leaves and flowers at the tops of the stems are colourful. Mostly yellows and reds. (Spring)
Nymania capensis (KLAPPERBOS)	The pink seed pods are most decorative. (Late spring and early summer)
Paranomus species (PARANOMUS, PERDEBOS)	Green, grey and mauvy-pink spikes. (Winter and early spring)
Metalasia species (METALASIA)	Rounded heads of flowers of yellow, mauve and ivory. Very long-lasting. (Winter and early spring)
Mundia spinosa (TORTOISE BERRY, DUINEBESSIE)	Graceful, slender stems of minute mauve flowers. (Late winter and early spring)
Phylica pubescens (FEATHERHEAD)	Green and yellowish-green terminal foliage. (Late winter)

NAME OF PLANT	MATERIAL
Podalyria calyptrata (KEURTJIE)	Has pink to mauve sweetly-scented, pea-shaped flowers in trusses. (Late winter and early spring)
Polygala virgata (PURPLE BROOM)	Slender spikes of deep purple flowers. (Spring)
Protea species (VARIOUS COMMON NAMES)	Bear heads which are large or of medium size. The colours are mostly pale to deep pink; ivory to sulphur-yellow; ice-green, russet and mahogany. (Winter to summer)
Serruria species (BLUSHING BRIDE AND OTHERS)	The serrurias are charming flowers for long-lasting arrangements—large or small, and for corsages and posies. Colours vary from alabaster to palest pink and rose. (Winter and early spring)
Stoebe plumosa (SLANGBOS)	Has grey stems of tiny grey leaves which last for weeks and are very effective in arrangements. (Spring to autumn)
Sutherlandia frutescens (CANCER BUSH, KANKERBOS)	Red flowers and glistening green seed pods. (Spring)

TREES FOR GARDENS AND PATIOS, AND FOR AVENUE OR STREET PLANTING

The owner of a new garden is always keen to plant trees as soon as possible to provide shade for the house and garden. Unfortunately, most of our indigenous trees are not very quick-growing, but this is no reason why some of them should not be planted in preference to exotic ones. One cannot change the nature of a plant, but very often trees are much slower in their development than they need be because they have been planted in poor soil or because they are not watered sufficiently during the dry months of the year. To promote good growth in regions where the soil is poor it is advisable to make holes 1 metre across and the same in depth, and to fill the holes with compost and manure. Full particulars of the trees listed below will be found in the text. Most of the trees listed as being suitable for large gardens are suitable also for avenue and street planting.

NAME OF PLANT	SMALL GARDEN	PATIO	LARGE GARDEN	DISTINCTIVE FOLIAGE	FLOWERS	EVERGREEN	DECIDUOUS	FODDER
Acacia albida (ANATREE, WHITETHORN, APIESDORING)			•	•	•		•	•
Acacia caffra (KAFFIRTHORN, KATDORING)	•		•	•	•		•	•
Acacia galpinii (MONKEYTHORN, APIESDORING)			•	•	•		•	•

NAME OF PLANT	SMALL GARDEN	PATIO	LARGE GARDEN	DISTINCTIVE FOLIAGE	FLOWERS	EVERGREEN	DECIDUOUS	FODDER
Acacia giraffae (Camelthorn, Kameeldoring)	•		•	•	•		•	•
Acacia haematoxylon (Vaalkameeldoring, Kaboom)	•	•		•			•	
Acacia karroo (Sweet-thorn, Mimosa, Soetdoring)	•	•	•	•	•		•	•
Acacia robusta (Enkeldoring, Oudoring)			•	•			•	
Acacia sieberiana var. **woodii** (Paperbarkthorn, Natal camelthorn)	•		•	•	•		•	•
Acacia tortilis subsp. heteracantha (Umbrellathorn, Tafelboom)	•		•	•	•		•	•
Acacia xanthophloea (Fevertree, Sulphur Bark)			•	•			•	
Afzelia cuanzensis (Rhodesian Mahogany)			•	•			•	
Alberta magna	•		•	•	•	•		
Albizia adianthifolia (Flat-crown, Munjerenje)	•		•	•	•		•	
Albizia harveyi (Platkroon)	•	•	•	•	•		•	
Apodytes dimidiata (White Pear, Witpeer)			•	•		•		
Baikiaea plurijuga (Rhodesian Teak, Rhodesian Chestnut)	•		•	•	•			
Bolusanthus speciosus (Tree Wistaria, Van Wykshout)	•		•		•		•	
Brabejum stellatifolium (Wild Almond, Wilde-amandel)	•					•		
Brachystegia spiciformis (Msasa)	•		•	•			•	

NAME OF PLANT	SMALL GARDEN	PATIO	LARGE GARDEN	DISTINCTIVE FOLIAGE	FLOWERS	EVERGREEN	DECIDUOUS	FODDER
Buddleia salviifolia (Sage Wood, Saliehout)	•	•	•		•	•		
Burkea africana (Wild Seringa, Rhodesian Ash)			•	•			•	
Calodendrum capense (Cape Chestnut, Wildekastaiing)	•		•	•	•		•	
Cassia abbreviata subsp. bearana (Long-tail cassia)	•	•	•	•	•		•	
Cassine crocea (Saffron, Safraan)			•	•		•		
Celtis africana (White stinkwood, Camdeboo stinkwood, Witstinkhout)			•	•			•	
Combretum erythrophyllum (Bush willow, Vaderlandswilg)	•		•				•	
Cunonia capensis (Red alder, Rooi-els)	•		•	•	•	•		
Curtisia dentata (Assegaaiwood, Assegaaihout)			•	•		•		
Cussonia paniculata and C. spicata (Cabbage tree, Kiepersol)	•	•	•	•		•		
Dais cotinifolia (Dais, Pompon Tree, Kannabas)	•	•	•	•	•	•	•	
Diospyros whytei (Black bark, Swartbas)	•	•	•	•		•		
Dodonaea viscosa (Sand olive, Sandolyf)	•	•				•		
Dombeya rotundifolia (Dombeya, Wild pear, Drolpeer)	•	•	•		•		•	
Ekebergia capensis (Cape ash, Essenhout)			•	•		•	•	
Ekebergia meyeri (Dog plum)	•		•	•		•		

NAME OF PLANT	SMALL GARDEN	PATIO	LARGE GARDEN	DISTINCTIVE FOLIAGE	FLOWERS	EVERGREEN	DECIDUOUS	FODDER
Erythrina caffra (Kaffirboom)	•		•		•		•	
Erythrina latissima (Broad-leaf kaffirboom)	•		•		•		•	
Erythrina lysistemon (Kaffirboom)	•		•		•		•	
Euclea crispa (Gwarri)	•					•		
Faurea macnaughtonii (Terblans)			•	•	•	•		
Gardenia spatulifolia (Transvaal gardenia)	•	•	•	•	•	•		
Gardenia thunbergia (Wild gardenia, Wilde katjiepiering)	•	•	•	•	•	•		
Gonioma kamassi (Kamassie, Knysna boxwood)	•	•	•	•	•	•		
Greyia sutherlandii (Natal bottlebrush, Baakhout)	•	•	•		•		•	
Halleria lucida (Tree Fuchsia, Witolyf, Notsung)	•	•			•	•		
Harpephyllum caffrum (Kaffirplum)			•	•		•		
Ilex mitis (Cape holly, Waterboom)	•		•	•		•		
Khaya nyasica (Red mahogany)			•	•		•		
Kigelia africana (Sausage tree, Cucumber tree)			•	•	•	•	•	
Kiggelaria africana (Wild peach, Wildeperske)	•		•			•		
Kirkia acuminata (White seringa, Witsering)			•	•			•	

NAME OF PLANT	SMALL GARDEN	PATIO	LARGE GARDEN	DISTINCTIVE FOLIAGE	FLOWERS	EVERGREEN	DECIDUOUS	FODDER
Kirkia wilmsii (Mountain seringa, Wild peppertree)	●		●	●			●	
Leucadendron argenteum (Silver tree)	●	●	●	●	●	●		
Loxostylis alata (Tierhout)	●	●	●	●		●		
Millettia grandis (Umzimbeet, Kaffir ironwood)			●	●		●	●	
Mimusops species (Milkwood trees or moepel)	●		●	●		●		
Nuxia floribunda (Vlier, White elder)	●		●		●	●		
Ochna pulchra (Ochna, Lekkerbreek)	●	●	●	●	●		●	
Olea africana (Wild olive, Olienhout)	●	●	●			●		●
Olea capensis subsp. **macrocarpa**			●	●		●		
Olinia cymosa (Hard pear, Rooibessie)			●	●		●		
Peltophorum africanum (African wattle, Rhodesian black wattle)	●		●	●	●		●	
Podalyria calyptrata (Shrub or tree, Keurtjie)	●	●			●	●		
Podocarpus henkelii (Henkel's yellowwood)			●	●		●		
Prunus africanum (Red stinkwood, Bitter almond)			●	●		●		
Pterocarpus angolensis (Kiaat, Transvaal teak)			●	●			●	
Pterocarpus rotundifolius (Roundleaf Kiaat)	●						●	

NAME OF PLANT	SMALL GARDEN	PATIO	LARGE GARDEN	DISTINCTIVE FOLIAGE	FLOWERS	EVERGREEN	DECIDUOUS	FODDER
Rapanea melanophloeos (Cape beech, Boekenhout)	•		•	•		•		
Rauvolfia caffra (Quinine tree, Nchongo)			•	•		•		
Rhus chirindensis (Red currant, Bostaaibos)			•	•			•	
Rhus lancea (Karee, Bastard Willow)	•	•	•			•		•
Salix capensis (Cape willow, Wild willow)	•		•			•	•	
Schotia brachypetala (Weeping boerboon, Tree fuchsia)	•		•		•		•	•
Sparrmannia africana (Stock rose)	•	•	•	•	•	•		
Spathodea campanulata (Rhodesian flame tree)	•	•	•	•	•	•	•	
Syzgium gerrardii (Water pear)			•			•		
Tamarix austro-africana (Wild tamarisk, Abiekwas)	•	•			•			•
Trema orientalis (Pigeon wood, Hophout)	•		•	•				
Trichilia roka (Natal or Cape mahogany)			•	•		•		
Trimeria rotundifolia (Wild Mulberry, Wildemoerbei)	•		•	•		•	•	
Vepris undulata (White ironwood, Witysterhout)			•	•		•		
Virgilia divaricata and **V. oroboides** (Keurboom)	•	•	•	•	•	•		
Ziziphus mucronata (Buffalo-thorn, Blinkblaar wag-'n-bietjie)	•		•	•		•	•	

COLOUR THROUGH THE SEASONS

In order to site shrubs and trees to best advantage in the garden it is important to know when they flower and the colours of the flowers, so that they may be planted where one colour will complement another. An effective and artistic grouping of plants can be achieved only if one knows their flowering time and colours. When planning the garden aim at achieving balance in the distribution of colour; bear in mind the seasons when plants flower so that you do not end up with a garden in which all the colour appears in one season leaving the garden rather drab for the rest of the year.

An attempt to classify flowers according to eight basic colours—using yellow, orange, red, pink, blue, mauve, purple and white—posed many problems because of the infinite number of shades in each of these. Red, for example, includes scarlet, crimson, tomato, coral and flame. Because of the fine gradations from one shade to another one can give only an indication of colour in a chart of this kind. For example, *Leucospermum reflexum,* which is a pretty shade of coral-flame is therefore listed under red and pink, although it is neither red nor pink.

Where a plant is listed under two or more colours it does not always mean that it bears flowers of different colours but that its flowers may be made up of shades of these. The colours of the flowers are more exactly described in the text.

Sometimes a plant is mentioned as flowering in two different seasons. This does not signify that it has a particularly long flowering time. It means that the plant may start flowering in one season and continue into the next, or that in one part of the country it may flower towards the end of a season whereas in another part it flowers at the beginning of the following season.

SHRUBS — SPRING

NAME OF PLANT	HEIGHT	YELLOW	ORANGE	RED	PINK	BLUE	MAUVE	PURPLE	WHITE
Acokanthera oppositifolia (Bushman's poison bush, Gifboom)	2–3 m				•				•
Adenandra species (China flower, Shepherd's delight)	20–60 cm				•				•
Agathosma crenulata (Oval leaf buchu)	1.5 m				•				•
Asclepias species (Wild cotton, Wildekapok)	2 m	•							•
Aspalathus species (South African gorse)	60 cm–2 m	•							
Aster filifolius (Wild Aster)	30–45 cm						•		
Athanasia species (Coulter bush, Koulterbos)	1-3 m	•							
Aulax species (male) (Aulax)	2 m	•							

NAME OF PLANT		YELLOW	ORANGE	RED	PINK	BLUE	MAUVE	PURPLE	WHITE
Bauhinia galpinii (PRIDE OF DE KAAP)	2 m			•					
Berkheya species (PRICKLY SUNFLOWER, DORINGGOUSBLOM)	60–90 cm	•							
Berzelia lanuginosa (KOLKOL)	2 m								•
Brunia species (BRUNIA, STOMPIE)	1–2 m								•
Burchellia bubalina (WILD POMEGRANATE, WILDEGRANAAT)	2–3 m			•					
Cadaba aphylla (DESERT BROOM, SWARTSTOOM)	1 m			•					
Carissa species (AMATUNGULU, NUM-NUM, NATAL PLUM)	1–3 m								•
Chrysanthemoides monilifera (BOETABESSIE)	2–2.5 m	•							
Clematopsis scabiosifolia (BUSH CLEMATIS, SHOCK-HEADED PETER, PLUIMBOSSIE)	1–1.5 m						•		
Combretum species (BURNING BUSH, RUSSET BUSH WILLOW, HICCUP NUT)	3–4 m				•				
Coleonema species (CONFETTI BUSH)	1–1.5 m				•				•
Crotalaria species (CAPE LABURNUM and others)	2–3 m	•					•		
Cyclopia genistoides (BUSH TEA, HONEY TEA)	1.5 m	•							
Erica ampullacea (BOTTLE HEATH, SISSIE HEATH)	60 cm				•				•
Erica baccans (BERRY HEATH)	1.5 m				•				
Erica bauera (ALBERTINIA HEATH, BRIDAL HEATH)	1 m				•				•

NAME OF PLANT	HEIGHT	YELLOW	ORANGE	RED	PINK	BLUE	MAUVE	PURPLE	WHITE
Erica blandfordia (Blandford's heath)	1 m	●							
Erica blenna (Lantern heath, Riversdale heath, Orange heath)	60 cm		●						
Erica campanularis (Yellow bell heath)	45 cm	●							
Erica cerinthoides (Red hairy erica, Rooihartjie)	30–90 cm			●					●
Erica chamissonis (Grahamstown heath)	60 cm				●				
Erica corifolia	45 cm				●				
Erica daphniflora (Daphne erica)	60 cm	●			●				●
Erica deliciosa (Port Elizabeth heath)	1–1.25 m				●				●
Erica densifolia (Knysna heath)	1 m			●					●
Erica eugenea	1 m				●				
Erica formosa	60 cm								●
Erica glauca (Cup and saucer heath)	1 m			●					
Erica glauca var. **elegans** (Petticoat heath)	1 m				●				
Erica grandiflora (Large orange heath)	1.5 m		●						
Erica hibbertia	60 cm			●					
Erica holosericea (Small flounced heath)	1 m				●				
Erica imbricata	1 m				●				●
Erica inflata	1 m				●				

NAME OF PLANT	HEIGHT	YELLOW	ORANGE	RED	PINK	BLUE	MAUVE	PURPLE	WHITE
Erica irregularis (Gansbaai heath)	1.5 m				•				
Erica nana (Dwarf heath)	30 cm	•							
Erica ovina (Woolly erica)	1 m								•
Erica ovina var. **purpurea** (Woolly erica)	1 m				•				
Erica perspicua (Prince of Wales heath)	1 m				•		•		
Erica peziza (Velvet bell heath, Kapokkie)	1.5 m								•
Erica quadrangularis (Pink shower heath)	45 cm				•				•
Erica regia (Royal heath, Elim heath)	60–90 cm			•					•
Erica thunbergii (Malay heath)	30–40 cm	•	•						
Erica ventricosa (Wax heath, Washeide, Franschhoek heath)	1 m				•				•
Erica vestita (Wide-mouthed heath)	1 m			•	•				•
Erica walkeria (Walker's heath, Swellendam heath)	60 cm				•				•
Erythrina acanthocarpa (Tamboekie thorn)	1–1.25 m			•					
Erythrina zeyheri (Prickly cardinal, Ploegbreker)	60–90 cm			•					
Euryops species (Daisy bush, Resin bush)	60–100 cm	•							
Gardenia amoena (East London gardenia)	2 m								•

NAME OF PLANT	HEIGHT	YELLOW	ORANGE	RED	PINK	BLUE	MAUVE	PURPLE	WHITE
Hermannia species (Dikbos, Mustard bush)	60–90 cm	●			●				
Hypericum revolutum (Curry bush, Forest primrose)	2–2.5 m	●							
Jasminum angulare (East London jasmine)	1.5 m +								●
Jasminum breviflorum (Natal jasmine)	2 m +								●
Jasminum multipartitum (Jasmine)	2 m +								●
Lebeckia cytisoides (Wild broom)	1.5 m	●							
Lebeckia simsiana (Dwarf lebeckia)	30 cm	●							
Leucadendron species (Leucadendron)	60 cm–2.5 m	●		●	●				
Leucospermum catherinae (Catherine wheel)	1.5 m	●			●				
Leucospermum conocarpodendron (Kreupelhout)	2–5 m	●							
Leucospermum cordifolium (Pincushion, Speldekussing)	1–1.25 m				●				
Leucospermum cuneiforme	1.5 m	●	●	●					
Leucospermum glabrum	1.25 m				●				
Leucospermum grandiflorum (Rainbow pincushion)	2 m	●	●	●					
Leucospermum gueinzii	1.5 m	●	●	●					
Leucospermum lineare (Narrow-leaf pincushion)	1.25 m				●				
Leucospermum muirii (Muir's pincushion)	1 m	●		●	●				

NAME OF PLANT	HEIGHT	YELLOW	ORANGE	RED	PINK	BLUE	MAUVE	PURPLE	WHITE
Leucospermum oleifolium (Tufted pincushion)	1–1.5 m	•	•	•					
Leucospermum prostratum (Creeping pincushion)	10 cm	•	•		•				
Leucospermum reflexum (Rocket pincushion)	2–3 m			•	•				
Leucospermum tottum	1 m				•				
Leucospermum truncatulum	1 m	•			•				
Leucospermum vestitum (Upright pincushion)	1.25 m		•	•					
Liparia splendens (Mountain dahlia, Geelkoppie)	1 m	•	and brown						
Lobostemon trigonous	30 cm				•	•			
Mackaya bella (Mackaya)	1.5–2 m						•		
Metalasia species (Metalasia)	30 cm–1.5 m	•					•		•
Mimetes species (Mimetes, soldaat)	1–2 m	•		•	•				
Nymania capensis (Nymania, Chinese lantern, Klapperbos)	2 m				•				
Oldenburgia arbuscula (Kreupelboom)	2–3 m						•		
Phygelius capensis (River bells)	1 m			•	•				
Podalyria calyptrata (Keurtjie)	2–4 m						•		•
Polygala virgata (Purple broom)	2 m							•	
Protea aristata (Ladismith protea)	1–2 m				•				

NAME OF PLANT	HEIGHT	YELLOW	ORANGE	RDE	PINK	BLUE	MAUVE	PURPLE	WHITE
Protea barbigera (GIANT WOOLLY PROTEA)	1.25 m	•			•				
Protea cedromontana (CEDARBERG PROTEA)	1 m						rust		
Protea compacta (BOT RIVER PROTEA)	2–3 m				•				
Protea cynaroides (GIANT OR KING PROTEA)	1–1.25 m				•				
Protea eximia (RAY-FLOWERED PROTEA)	2 m				•				
Protea grandiceps (PEACH PROTEA)	1–1.5 m				•				
Protea lacticolor (BABY PROTEA)	2–3 m				•				
Protea laurifolia (FRINGE PROTEA)	2.5 m				•				
Protea lepidocarpodendron (BLACK-BEARDED PROTEA)	2 m	•	and green						
Protea longiflora (LONG-BUD PROTEA)	2 m	•			•				
Protea longifolia (LONG-LEAFED PROTEA)	1–1.5 m	•							
Protea nana (MOUNTAIN ROSE, BERGROOS, SKAAMROOS)	60–90 cm				•	and rust			
Protea obtusifolia (BREDASDORP PROTEA)	2–3 m				•	and green			
Protea pityphylla (PINE-LEAFED PROTEA)	1 m				rust				
Protea scolymocephala (SMALL GREEN PROTEA)	1 m	green							
Protea speciosa (BROWN-BEARDED PROTEA)	1.25 m	•			•				

NAME OF PLANT	HEIGHT	YELLOW	ORANGE	RED	PINK	BLUE	MAUVE	PURPLE	WHITE
Rothmannia capensis (Scented cups)	3 m								•
Rothmannia globosa (September bells)	5 m								•
Serruria species (Blushing bride, Grey serruria, Silky serruria, Spider bush)	1 m				•		and grey		
Sutherlandia frutescens (Cancer bush, Kankerbos, Gansies)	1–2 m			•					
Thunbergia alata (Black-eyed susan)	climber	•	•						

TREES

NAME OF PLANT	HEIGHT	YELLOW	ORANGE	RED	PINK	BLUE	MAUVE	PURPLE	WHITE
Acacia giraffae (Camelthorn, Kameeldoring)	8 m	•							
Acacia nigrescens (Knobthorn, Knoppiesdoring)	8–10 m	•							
Acacia sieberiana var. **woodii** (Paperbarkthorn, Natal camelthorn, Platkroonsoetdoring)	8 m	•							•
Acacia tortilis subsp. **heteracantha** (Umbrellathorn, Tafelboom, Haak-en-steek)	8 m	•							
Acacia xanthophloea (Fevertree, Sulphur bark, Koorsboom)	12 m	•							
Afzelia cuanzensis (Rhodesian mahogany, Mahogany bean)	15–18 m				•				
Alberta magna (Alberta)	3–6 m			•					

NAME OF PLANT	HEIGHT	YELLOW	ORANGE	RED	PINK	BLUE	MAUVE	PURPLE	WHITE
Albizia adianthifolia (Flat-crown, Platkroon, Munjerenje)	8 m	•							
Bolusanthus speciosus (Tree wistaria, Van Wykshout)	6 m						•		
Dais cotinifolia (Dais, Pompon tree, Kannabas)	5 m				•		•		
Freylinia lanceolata (Freylinia)	5 m	•							
Gardenia spatulifolia (Transvaal gardenia)	5 m								•
Greyia species	3–4 m			•					
Leucadendron argenteum (Silver tree)	5–8 m	silver							
Ochna pulchra (Ochna, Lekkerbreek)	5–6 m	•							
Peltophorum africanum (African wattle, Rhodesian black wattle, Huilboom)	10 m	•							
Pterocarpus angolensis (Kiaat, Transvaal teak)	10 m	•							
Rhus lancea (Karee, Bastard willow)	5-7 m				green				
Schotia brachypetala (Weeping boerboon, Huilboerboon, Tree fuchsia)	7–10 m			•					
Sparrmannia africana (Stock rose)	5–7 m								•
Spathodea campanulata (Rhodesian flame tree)	9–10 m			•					
Virgilia divaricata (Keurboom)	6–8 m						•		

SHRUBS

NAME OF PLANT	HEIGHT	YELLOW	ORANGE	RED	PINK	BLUE	MAUVE	PURPLE	WHITE
Acridocarpus natalitius (Feather climber)	2 m	•							
Asclepias species (Wild cotton, Wildekapok)	2 m	green pods							
Bauhinia galpinii (Pride of De Kaap)	2 m			•					
Burchellia bubalina (Wild pomegranate)	2–3 m			•					
Carissa species (Amatungulu)	1–3 m								•
Clematis brachiata (Traveller's joy, Old man's beard, Klimop)	1–1.5 m								•
Clematopsis scabiosifolia (Bush clematis, Shock-headed peter, Pluimbossie)	1–1.5 m						•		
Dissotis canescens (Wild lasiandra)	1 m						•	•	
Dombeya burgessiae (Pink-and-white dombeya)	2–3 m								•
Dombeya tiliacea (Heart-leaf dombeya)	2–3 m								•
Dovyalis caffra (Kei apple, Dingaan's apricot)	3 m	yellow fruit							
Duvernoia adhatodioides (Pistol bush)	2.5 m								•
Erica bergiana (Fairy erica)	1.5 m				•				
Erica borboniaefolia (Flounced heath)	60 cm				•				
Erica corifolia	45 cm				•				•

NAME OF PLANT	HEIGHT	YELLOW	ORANGE	RED	PINK	BLUE	MAUVE	PURPLE	WHITE
Erica curviflora (Water heath, Waterbos)	30–90 cm		●	●					
Erica curvirostris (Scented bell heath)	60 cm				●				●
Erica daphniflora (Daphne erica)	60 cm	●			●				●
Erica densifolia (Knysna heath)	1 m			●					●
Erica grandiflora (Large orange heath)	1.5 m		●						
Erica inflata	1 m				●				
Erica infundibuliformis (Funnel heath)	1 m				●				●
Erica junonia	30 cm				●				
Erica lateralis (Globe heath)	45 cm				●				
Erica longifolia (Long-leafed heath)	1 m	●	●	●	●			●	
Erica lutea	1 m	●							●
Erica mammosa (Red signal heath, Rooiklossie heide)	1 m		●	●	●				●
Erica massoni (Houhoek heath)	45 cm		●	●					
Erica parilis	60 cm	●							
Erica shannonea (Star-faced heath)	45 cm								●
Erica taxifolia (Double pink heath)	40–60 m				●				
Erica tumida (Mountain heath)	1 m			●					

NAME OF PLANT	HEIGHT	YELLOW	ORANGE	RED	PINK	BLUE	MAUVE	PURPLE	WHITE
Erica ventricosa (Wax heath, Washeide, Franschhoek heath)	1 m				•				•
Erica viridiflora (Green heath)	1 m	green							
Erica woodii (Wood's heath)	60 cm				•				•
Erythrina humeana (Dwarf kaffirboom)	2–3 m			•					
Euryops acraeus (Mountain daisy)	60–90 cm	•							
Grewia occidentalis (Assegai wood, Crossberry)	2 m				•				
Leucospermum muirii (Muir's pincushion)	1 m	•	•	•	•				
Oncoba spinosa (Oncoba)	3 m								•
Pavetta species (Christmas bush)	1–3 m								•
Phygelius capensis (River bells)	1 m			•	•				
Plumbago auriculata (Plumbago)	1–2 m					•			
Podranea brycei (Zimbabwe creeper)	climber				•				
Podranea ricasoliana (Port St John's climber)	climber				•				
Protea aristata (Ladismith protea)	1–2 m				•				
Protea caffra (Highveld protea)	4–5 m				•				
Protea cryophila (Snow protea)	30–45 cm								•

NAME OF PLANT	HEIGHT	YELLOW	ORANGE	RED	PINK	BLUE	MAUVE	PURPLE	WHITE
Protea lacticolor (Baby protea)	2–3 m				●				
Protea rouppelliae (Drakensberg protea)	3–4 m				●				
Rothmannia capensis (Scented cups)	3 m								●
Syncolostemon densiflorus (Pink plume)	1 m				●				
Tecomaria capensis (Cape honeysuckle)	3 m	●	●						
Thunbergia alata (Black-eyed susan)	climber	●	●						

TREES

NAME OF PLANT	HEIGHT	YELLOW	ORANGE	RED	PINK	BLUE	MAUVE	PURPLE	WHITE
Acacia caffra (Kaffirthorn, Katdoring)	8 m	●							
Acacia galpinii (Monkeythorn, Apiesdoring)	10–15 m	●							
Acacia karroo (Sweet-thorn, Mimosa, Soetdoring, Witdoring)	5 m	●							
Calodendrum capense (Cape chestnut)	8–10 m				●				
Ekebergia capensis (Cape ash, Essenhout, Dog plum)	10 m	●							
Gardenia spatulifolia (Transvaal gardenia)	5 m								●

NAME OF PLANT	HEIGHT	YELLOW	ORANGE	RED	PINK	BLUE	MAUVE	PURPLE	WHITE
Gardenia thunbergia (Wild gardenia, Wilde katjiepiering)	3 m								●
Leucadendron argenteum (Silver tree)	5–8 m	silver leaves							
Millettia grandis (Umzimbeet, Kaffir ironwood)	10 m							●	
Trichilia roka (Natal or Cape mahogany, Rooiessenhout)	10–12 m								●
Virgilia oroboides	6–8 m						●		

SHRUBS

NAME OF PLANT	HEIGHT	YELLOW	ORANGE	RED	PINK	BLUE	MAUVE	PURPLE	WHITE
Barleria obtusa (Barleria)	1–2 m						●		
Bauhinia glapinii (Pride of de kaap)	2 m			●					
Clematis brachiata (Traveller's joy, Old man's beard, Klimop)	1–1.5 m								●
Dissotis species	1–2 m						●	●	
Dombeya burgessiae (Pink-and-white dombeya)	2–3 m								●
Dombeya tiliacea (Heart-leaf Dombeya)	2–3 m								●
Duvernoia adhatodioides (Pistol bush)	2.5 m								●
Erica chloroloma (Green-fringed heath)	2.5 m			●					
Erica curvirostris (Scented bell heath)	60 cm				●				●
Erica fascicularis (Tigerhoek heath)	1.25 m				●				
Erica glandulosa (Sticky-leaved heath)	1.5 m		●		●				
Erica mammosa (Red signal heath, Rooiklossie heide)	1 m		●	●	●				●
Erica oatesii	1 m			●	●				
Erica pillansii (Pillans' heath)	1 m			●					
Erica woodii (Wood's heath)	60 cm				●				●
Leucadendron salignum (Geelbos)	60 cm	●							

NAME OF PLANT	HEIGHT	YELLOW	ORANGE	RED	PINK	BLUE	MAUVE	PURPLE	WHITE
Paranomus reflexus (Green paranomus)	1.25 m	green							
Paranomus spicatus (Perdebos)	1 m						•		
Plectranthus behrii (Pink spurflower)	60–90 cm				•				
Plectranthus ecklonii (Purple spurflower)	1–2 m							•	
Plumbago auriculata (Plumbago)	1–2 m					•			
Podranea brycei (Zimbabwe creeper)	climber				•				
Podranea ricasoliana (Port St John's climber)	climber				•				
Protea arborea (Waboom)	3–4 m	•							
Protea longifolia (Long-leafed protea)	1–1.5 m	•	green						
Protea macrocephala (Green protea)	2–3 m	green							
Protea minor (Ground rose, Aardroos)	30 cm				•	rust			
Protea neriifolia (Oleander-leafed protea, Blousuikerbos)	1.5–3 m	•			•	green			
Protea pulchra (Gleaming protea)	2 m	green			•				
Protea repens (Sugarbush, Suikerbos)	1.5–2 m	•			•	green			
Protea speciosa (Brown-bearded protea)	1.25 m	•			•				
Pycnostachys urticifolia (Blue boys, Porcupine salvia)	2 m					•			

NAME OF PLANT	HEIGHT	YELLOW	ORANGE	RED	PINK	BLUE	MAUVE	PURPLE	WHITE
Senecio tamoides (CANARY CREEPER)	climber	●							
Tecomaria capensis (CAPE HONEYSUCKLE)	3 m	●	●						

TREES

NAME OF PLANT	HEIGHT	YELLOW	ORANGE	RED	PINK	BLUE	MAUVE	PURPLE	WHITE
Combretum erythrophyllum (BUSH-WILLOW, VANDERLANDSWILG)	6–8 m	autumn foliage							
Cunonia capensis (RED ALDER, ROOI-ELS)	8–10 m								●
Kirkia acuminata and **K. wilmsii** (WHITE SERINGA, WITSERING)	8–10 m	autumn foliage							
Leucadendron argenteum (SILVER TREE)	5–8 m	silver leaves							

SHRUBS

NAME OF PLANT	HEIGHT	YELLOW	ORANGE	RED	PINK	BLUE	MAUVE	PURPLE	WHITE
Adenandra species (CHINA FLOWER, SHEPHERD'S DELIGHT)	20–60 cm				●				●
Agathosma crenulata (OVAL LEAF BUCHU)	1.5 m				●				●

NAME OF PLANT	HEIGHT	YELLOW	ORANGE	RED	PINK	BLUE	MAUVE	PURPLE	WHITE
Aster filifolius (Wild aster)	30–45 cm						•		
Coleonema species (Confetti bush)	1–1.5 m				•				•
Erica ampullacea (Bottle heath, Sissie heath)	60 cm				•				•
Erica aristata	60 cm				•				
Erica baccans (Berry heath)	1.5 m				•				
Erica blenna (Lantern heath, Riversdale heath, Orange heath)	60 cm		•						
Erica bodkinii	60 cm								•
Erica campanularis (Yellow bell heath)	45 cm	•							
Erica cerinthoides (Red hairy erica, Rooihartjie)	30–90 cm			•	•				
Erica chloroloma (Green-fringed heath)	2.5 m			•					
Erica deliciosa (Port Elizabeth heath)	1.25 m				•				•
Erica eugenea	1 m				•				
Erica fascicularis (Tigerhoek heath)	1.25 m				•				
Erica formosa	60 cm								•
Erica glandulosa (Sticky-leafed heath)	1.5 m		•		•				
Erica glauca (Cup-and-saucer heath)	1 m			•					
Erica glauca var. **elegans** (Petticoat heath)	1 m				•				

NAME OF PLANT	HEIGHT	YELLOW	ORANGE	RED	PINK	BLUE	MAUVE	PURPLE	WHITE
Erica imbricata	1 m				•				•
Erica irregularis (Gansbaai heath)	1.5 m				•				
Erica oatesii	1 m			•	•				
Erica patersonia (Mealie heath)	1 m	•							
Erica perspicua (Prince of Wales heath)	1 m				•		•		
Erica pillansii (Pillans' heath)	1 m			•					
Erica porteri	1 m			•					
Erica regia (Royal heath, Elim heath)	60–90 cm			•					•
Erica sessiliflora (Green heath)	1.25 m	green							
Erica thunbergii (Malay heath)	30–40 cm	•	•						
Erica walkeria (Walker's heath, Swellendam heath)	60 cm				•				•
Erythrina acanthocarpa (Tamboekie thorn)	1–1.25 m			•					
Euryops species (Daisy bush, Resin bush)	60–100 cm	•							
Gardenia amoena (East London gardenia)	2 m								•
Iboza riparia (Iboza)	2 m						•		
Lachnaea densiflora (Bergaster)	40 cm	•							•
Lebeckia cytisoides (Wild broom, Wilde besembos)	1.25 m	•							

NAME OF PLANT	HEIGHT	YELLOW	ORANGE	RED	PINK	BLUE	MAUVE	PURPLE	WHITE
Lebeckia simsiana (Dwarf lebeckia)	30 cm	•							
Leucadendron species (Leucadendron)	60 cm–2.5 m	•		•	•				
Leucospermum catherinae (Catherine wheel)	1.5 m	•			•				
Leucospermum conocarpodendron (Kreupelhout)	2–5 m	•							
Leucospermum cordifolium (Pincushion, Speldekussing)	1–1.25 m				•				
Leucospermum cuneiforme	1.5 m	•	•	•					
Leucospermum glabrum	1.25 m				•				
Leucospermum lineare (Narrow-leaf pincushion)	1.25 m				•				
Leucospermum muirii	1.25 m	•	•						
Leucospermum reflexum (Rocket pincushion)	2–3 m			•	•				
Leucospermum vestitum (Upright pincushion)	1.25 m		•	•					
Liparia splendens (Mountain dahlia, Geelkoppie)	1 m	•	and brown						
Lobostemon fruticosus (Eight-day-healing bush)	1 m				•	•			
Metalasia aurea (Metalasia)	45 cm	•							
Metalasia muricata (Blombos)	1.5 m								•
Metalasia seriphiifolia (Mauve metalasia)	1.5 m						•		
Mimetes cucullatus (Mimetes, Soldaat)	1.5 m	•		•					

NAME OF PLANT	HEIGHT	YELLOW	ORANGE	RED	PINK	BLUE	MAUVE	PURPLE	WHITE
Mimetes hirtus	1 m	●		●	●				
Mundia spinosa	1.25 m						●		
Paranomus reflexus (Green paranomus)	1.25 m	green							
Paranomus spicatus (Perde bos)	1 m						●		
Podalyria calyptrata (Keurtjie)	2–4 m				●		●		●
Polygala myrtifolia (Septemberbossie)	1–1.5 m						●		
Polygala virgata (Purple broom)	2 m							●	
Priestleya villosa (Silver pea)	1.25 m	●							
Protea arborea (Waboom)	3–4 m	●							
Protea barbigera (Giant woolly protea, Queen protea)	1.25 m	●			●				
Protea cedromontana (Cedarberg protea)	1 m				rust				
Protea compacta (Bot River protea)	2–3 m				●				
Protea cynaroides (Giant or King protea)	1–1.25 m				●				
Protea eximia (Ray-flowered protea)	2 m				●				
Protea grandiceps (Peach protea)	1–1.5 m			●					
Protea lepidocarpodendron (Black-bearded protea)	2 m	●	green						
Protea longiflora (Long-bud protea)	2 m	●			●				

NAME OF PLANT	HEIGHT	YELLOW	ORANGE	RED	PINK	BLUE	MAUVE	PURPLE	WHITE
Protea macrocephala (Green protea)	2–3 m	green							
Protea minor (Ground rose, Aardroos)	30 cm				•	rust			
Protea nana (Mountain rose, Bergroos, Skaamroos)	60–90 cm				•	rust			
Protea neriifolia (Oleander-leafed protea, Blousuikerbos)	1.5–3 m	•			•	green			
Protea obtusifolia (Bredasdorp protea)	2–3 m				•	green			
Protea pityphylla (Pine-leafed protea)	1 m	rust							
Protea pulchra (Gleaming protea)	2 m	green			•				
Protea repens (Sugarbush, Suikerbos)	1.5–2 m	•			•	green			
Protea scolymocephala (Small green protea)	1 m	green							
Protea speciosa (Brown-bearded protea)	1.25 m	•			•				
Protea stokoei (Stokoe's protea)	1–1.5 m				•				
Protea susannae (Susan's protea)	1–1.5				•				
Pycnostachys urticifolia (Blue boys, Porcupine salvia)	2 m					•			
Rafnia ovata (Rafnia)	2 m	•							
Rafnia thunbergii	1 m	•							
Serruria species (Blushing bride, Grey serruria, Silky serruria, Spider bush)	1 m				•	and grey			•

TREES

NAME OF PLANT	HEIGHT	YELLOW	ORANGE	RED	PINK	BLUE	MAUVE	PURPLE	WHITE
Acacia albida (Anatree, Anaboom, Whitethorn)	18 m	●							
Acacia giraffae (Camelthorn, Kameeldoring)	8 m	●							
Acacia nigrescens (Knobthorn, Knoppiesdoring)	8–10 m	●							
Alberta magna (Alberta)	3–5 m			●					
Buddleia salviifolia (Sage wood, Saliehout)	6 m						●		
Dombeya rotundifolia (Wild pear, Dombeya, Dikbas)	5 m								●
Erythrina caffra (Kaffirboom, Lucky bean tree)	10 m			●					
Erythrina latissima (Broad-leaf kaffirboom)	6 m			●					
Erythrina lysistemon (Kaffirboom, Lucky bean tree)	8 m			●					
Greyia sutherlandii (Natal bottlebrush, Baakhout, Mountain bottlebrush)	3 m			●					
Kigelia africana (Sausage tree, Cucumber tree)	10 m			●					
Leucadendron argenteum (Silver tree)	5–8 m	silver leaves							
Nuxia floribunda (Vlier, White elder)	8–12 m								●

III

Part III
Shrubs

◁ The flowers of *Leucospermum tottum* are lovely in form and colouring.

Shrubs

As we become more and more involved in a great diversity of activities we have less and less time to devote to beautifying our home surroundings, with the result that most of us now aspire to have gardens which are pleasant in appearance throughout the year without demanding much attention. Gardens of today are therefore quite different from those of a generation ago. Then, there were large and ornate beds of annuals and herbaceous borders; today the garden is dominated by perennial plants, and particularly by shrubs and trees, planted to create a beautiful setting that requires little labour to keep it attractive.

All too often gardeners, impatient to see results, plant without planning and end up with an unsightly jungle instead of a beautiful garden. As shrubs are permanent it is advisable to give a good deal of thought to choosing suitable varieties, and to deciding on their position. Because you have been captivated by the beauty of a shrub do not order it from a nursery until you have decided where to plant it as, to look effective, it must form part of the total scene.

Before planting a shrub it is advisable also to learn something about its likes and dislikes; what soil conditions suit it; whether it prefers sunshine or shade and whether it stands severe frost or does best in a mild climate. Above all, try to form a complete picture in your mind, not of any single plant which may have fascinated you, but of the whole garden. Plan where you can put the shrub so that it not only shows up to best advantage itself, but so that it will blend in happily with other plants in that part of the garden where you intend placing it.

In South Africa our native shrubs have been neglected for years. In fact it is only within the last decade that gardeners in South Africa have begun to use indigenous shrubs in their gardens. This is strange, as gardeners in Europe, where conditions are not suitable for most of them, have been growing some of them for generations, and, in Australia and New Zealand, many of our native shrubs are far better known and better grown than they are here, in their homeland. It is not only gardeners who have failed to appreciate the beauty of South African plants; florists too, have shown little interest in native plant material. Until only a few years ago they preferred making arrangements and bouquets of the many exotic plants which grow so well in Southern Africa, and their arrangements invariably included sweet peas, carnations, roses, gladioli, daffodils and so on. Now, however, the picture has changed and in some parts of South Africa 75% of the flowers used, particularly in winter and spring, are native plants, more particularly members of the protea family and heaths. This is not surprising as some of these plants are not only extremely decorative but they are also long-lasting when picked.

CHOOSING SOUTH AFRICAN SHRUBS FOR YOUR GARDEN

Before choosing shrubs for your garden the following three factors should be carefully considered:

(*a*) **Climate:**

Choose shrubs to suit the climate. Shrubs are adaptable but, if you live in a district which experiences extremes, it may be difficult to rear some of the plants selected. If, for example, you live in an area where growing conditions are difficult because of intensely hot weather, little rain and a shortage of water for irrigation, there is no point in trying to grow shrubs which need a good deal of water. Rather choose plants which grow naturally in regions where hot dry periods are usual. There are many parts of Southern Africa, Australia, South America, California and some of the southern states of the U.S.A., such as Arizona, Texas and New Mexico, where intense heat, dry air and a shortage of water make it difficult to garden with plants other than those adapted by nature to tolerate such conditions. There is no reason, however, why gardens in such areas should be drab because there are many South African native plants which will, in the words of Isaiah, make the desert "blossom as the rose".

On the other hand, gardeners who live in areas where the rainfall is high should choose, for the most part, plants which like wet conditions rather than those which flourish naturally in dry places; those who live in regions which experience severe frost should limit their gardening to the growing of plants which can survive such cold or else be prepared to protect the plants in winter; and gardeners near the sea, where the soil is often a porous type of sand, should choose mainly plants which do not mind this kind of soil. Furnishing your garden with plants which suit your climate is the first step towards having a labour-saving garden.

(*b*) **Size:**

The next point to consider when choosing shrubs for the garden is the size which the plants are likely to attain in your area. In suburban gardens of average size where space is limited, it is advisable to plant only shrubs which are in scale with the plot. Many proteas, for example, are too large to use effectively in small suburban gardens, but there are other shrubs of smaller growth which may be planted. It is true that many plants can be cut back to moderate size, but, if they have to be cut back more than once a year to keep them within bounds, they are unlikely to flower well.

(*c*) **Colour through the Seasons:**

In gardening one aims at having some colour in the garden throughout the year, and fortunately, in the mild climate which characterises a good deal of the Southern Hemisphere and parts of the Northern Hemisphere, this is not difficult.

Before choosing shrubs make a note of the season of the year when they flower. This will enable you to group them so that you have colour throughout the year and not all at one season. It will also enable you to plant them so that all the colour does not occur at one point in the garden at the same time. One should, for example, avoid having ALL the winter-flowering or spring-flowering shrubs together, as an abundance of colour on one side and none on the other tends to make the garden look unbalanced.

Knowing when different shrubs flower AND the colours of the flowers will enable you to make an effective grouping. Although colours do not clash outside as they would indoors, there is no doubt that to all of us some colours blend more harmoniously than do others. Most of us are offended by a close grouping of strong colours such as bright orange or red with cerise, whereas we are soothed and enchanted by a colour grouping which embraces pastel shades of blue and pink or even strong shades of these two colours, or blue and yellow, but not pink and yellow. Yellow and pink do not clash, but yellow looks far lovelier when combined with white, if it is a vivid yellow, and with other colours such as lilac to lavender and any shade of blue or orange, if it is a pale yellow.

When planting shrubs set two or more which flower at the same time near each other so that the colours enhance each other. For example, if you plant plumbago to give you colour during the hottest months of summer, then behind the plumbago plant a shrub such as the Cape honeysuckle (*Tecomaria capensis* var. *lutea*) which flowers at the same time, so that its sulphur-yellow flowers show up the blue of the plumbago.

When choosing shrubs make a note of the foliage of the plants too, as the colour of the leaves and their general appearance contributes as much, or more, to the garden than the flowers they bear. There is tremendous variation in shades of green, and there are some most decorative shrubs which have leaves with a tinge of grey or silver. Plants with grey leaves are invaluable in the garden in creating contrast and emphasis, and often they are most useful, also, in providing material for flower arrangements.

PREPARING THE GROUND FOR SHRUBS

Many South African shrubs grow readily in poor soil, but most of them do better when planted in soil to which compost or leafmould has been added. Some plants, such as proteas and ericas, thrive only in acid soil, and details of the kind of soil preparation necessary for them are given in the pages dealing with these particular plants. If the soil in your garden is not good, dig holes 60 cm (2 ft) deep and as much across and put in two or three shovelsful of compost, manure or leafmould. Then put back some of the soil previously removed and, with a fork, mix this up together. Fill the hole to within 25 cm (10 in) of the surface with this mixture and water it well or tramp it down to settle the soil. If fresh manure has been used, add a layer of soil about 5 cm (2 in) thick over the mixture in the hole as fresh manure may burn the roots of newly-set-out plants.

PLANTING:

When transplanting a young shrub from its container try to get it out with a ball of soil still adhering around the roots and plant it at the same depth as it stood in the container. Many shrubs are now potted in plastic containers and cutting away the plastic is a simple operation, but, if the container is a tin, removing the plant with earth around its roots is not always easy. Water the soil in the tin well beforehand and then tap the tin on something hard, turning it all the time. This will usually loosen the soil so that the plant comes out of the container with the soil adhering to its roots in a block.

Evergreen shrubs are always sent out from nurseries in containers, but deciduous ones may be sent out in winter with bare roots wrapped in moss and hessian only, or they may be sent out planted in containers at any season of the year.

Plants in containers may be planted out at any time of the year but generally it is easier to get them established if they are planted out just before or during the rainy season. In Southern Africa this means planting them out from October to January in the summer-rainfall region, and in March to August in the winter-rainfall region. As a rule it is best to plant out deciduous shrubs in late winter before they form new leaves. The time of the year to transplant is not vitally important, however, if one is prepared to care for the plants and to water them adequately until the roots are established. In areas where the air is very dry, it is advisable to spray the leaves of plants recently set out, as well as making sure that the ground is watered thoroughly.

PRUNING:

The pruning of shrubs is not always necessary and it is not as important in producing good results as it is in the case of roses and fruit trees where, if pruning is not done, the quality of the flowers or fruit tends to be poor. With shrubs pruning is done

(i) to keep them from growing too large for their allotted space in the garden, and

(ii) to keep them shapely and to force new growth. Some plants become woody and leggy as they grow older and these should be pruned back lightly each year to encourage them to make new growth from the base. The best time to do this cutting back is soon after they have flowered. Often merely cutting off faded flowers with a long piece of stem is sufficient. How much to cut off depends upon the size and vigour of the plant. Err on the side of light pruning rather than heavy cutting back until you know more about the plant's growth, and, if you are not certain what to do about young plants, leave them to grow on until you have acquired a more intimate knowledge of their character.

PROPAGATING SHRUBS:

Most gardeners purchase shrubs from nurseries and this is the quickest and easiest way to establish your garden. If you feel, however, that you would like to propagate some shrubs on your own, try growing them from seed or from cuttings 7-15 cm (3-6 in) long. Generally it can be said that

The Feather Climber *(Acridocarpus natalitius)* has decorative leaves and flowers.

growing shrubs from seed takes longer than growing them from cuttings, and that the seed of most shrubs takes longer to germinate than does the seed of annuals and perennials.

Cuttings of plants which are deciduous are usually made of hard or mature wood in early autumn, whilst those of evergreen plants are made in early spring, of soft wood, but not the very soft ends of stems, except in certain cases, such as ericas.

ACOKANTHERA OPPOSITIFOLIA

(A. venenata) BUSHMAN'S POISON BUSH, GIFBOOM

DISTRIBUTION: Is found in the eastern Cape from the Port Elizabeth area into Natal, usually at the edges of forests near the coast.

DESCRIPTION: Although this is a handsome shrub it cannot be recommended for the garden unless care is taken to remove the flowers as they fade and before they form fruits, for the fruit which looks like small plums is extremely poisonous. It is as well to learn to recognise the plant because all parts of it are poisonous, and even the use of twigs as skewers on which to grill meat could be fatal. The Bushmen used this plant to provide poison with which to tip their arrows. Sims, who did much research and writing on our native plants made the following observation . . . "With its attractive flowers, foliage and berries, it is perhaps the most dangerous poison our forests contain".

It is a handsome evergreen shrub growing to about 2-3 m (6-10 ft) or more. The leaves are dark green, glossy, leathery and oval with a pointed apex. The ivory or white flowers tinged with pink appear in clusters about 5-7 cm across. Each little flower consists of a slender tube often flushed with pink, opening to a starry five-petalled face about 1 cm across. They appear in spring and are very attractive, and sweetly-scented. The fruit which is the size of small plums ripens to a purplish-black colour.

CULTURE: This shrub is tender to frost and needs good soil, plenty of moisture and some shade to promote good growth.

ACRIDOCARPUS NATALITIUS FEATHER CLIMBER

DISTRIBUTION: Occurs in the eastern Cape, Natal and in the lowveld of the Transvaal.

DESCRIPTION: This is a scandent shrub which grows to about 2 m (6 feet) in height, and has dark shiny leaves which are oval and pointed, much longer than they are broad. In summer it bears conical spikes of flowers, each flower being 2-3 cm across. They are made up of five bright yellow petals rather crinkled at the outside edges. This is a pretty plant for the shrub border, or to have against a wall.

CULTURE: It grows well in full sun near the coast but should be planted where it is shaded from the heat of the day when grown inland. It needs regular watering and plenty of compost in the soil. It can be grown from seed or cuttings. It stands only moderate frost.

ADENANDRA CHINA FLOWER, SHEPHERD'S DELIGHT

DISTRIBUTION: These are to be found in a limited area in the western and south-western Cape.

DESCRIPTION: The plants reach a height of between 25 and 60 cm (10-24 in), depending on the species. The leaves are sometimes greyish-green on the underside, very small and aromatic. The flowers look as though they have been fashioned from porcelain which accounts for the common name of "china flower". They have five petals and measure 1-3 cm across. The petals are white or palest shell-pink sometimes with a rose coloured stripe down the centre; the buds are flushed with rose and so are the backs of the petals of the open flower. Adenandra flowers in early spring. The species mentioned below look effective in a tub on a patio, in the garden amongst flowers, in front of the shrub border or in a rock-garden.

CULTURE: They should be watered regularly particularly in autumn and winter to encourage good flowering. China flower stands a good deal of frost, but in hot inland districts it should be planted so that it is shaded by other plants during the hottest hours of the day. It can be grown from seed sown in spring or from tip cuttings made in spring or early summer.

A. fragrans DWARF CHINA FLOWER

Grows to only about 30 cm (1 ft) and bears beautiful little pink enamelled flowers in September. The leaves are aromatic and very small, scarcely a centimetre in length. This is a charming plant for pots or small gardens. It flowers in late winter and early spring.

A. serphyllacea

Is a very small shrublet suitable for the small garden or rockery. Its flowers of palest pink scintillate in the sunlight.

A. umbellata SHEPHERD'S DELIGHT

(*A. cuspidata*)

It has been suggested that the charming common name was given because the flowering season of adenandra heralds the onset of spring when good grazing can be expected. This species grows to about 60 cm (2 ft) in height and spread. Its leaves are often larger than those of the other two

Adenandra serphyllacea. This species is a charming plant for the small rock-garden.

species described. The flowers, which are about 18 mm across, vary in colour from white to pale pink. In September, when it is in full flower, it is a most decorative plant.

A. uniflora CHINA FLOWER, KOMMETJIETEEWATER

Reaches a height of 30 cm (1 ft) and has little, oval pointed leaves about 1 cm in length. The flowers are ivory to palest pink with deep rose lines down the middle. The flowering time is from July to October.

AGATHOSMA CRENULATA OVAL LEAF BUCHU
(Barosma crenulata)

DISTRIBUTION: Occurs in many parts of the south-western Cape and north of Cape Town.

DESCRIPTION: It grows to a height of 1.5 m (4 ft) with a spread of as much. The leaves are oval and only about 2 cm in length. The individual flowers are very small and not showy, but they appear in such quantities that the shrub is attractive when in full flower in late winter and early spring. Each flower is star-shaped with five petals which are generally white but which may be palest pink. The leaves are aromatic and were used by the Hottentots in the treatment of various ills. Later, settlers also used infusions of this plant with brandy or vinegar to cure ailments, principally of the stomach.

CULTURE: This plant does not mind poor soil but flowers better when grown in soil to which a little compost has been added. It endures quite severe frost and once established will tolerate long periods of drought. It is easily propagated from seeds or cuttings.

ASCLEPIAS WILD COTTON, WILDE KAPOK

DISTRIBUTION: There are over fifty species in different parts of Southern Africa, many of them in the Transvaal and Rhodesia.

DESCRIPTION: This plant, which grows so easily and quickly that it might be regarded as a weed, is a useful one for providing something green in dry places, and it also produces seed pods which look effective in arrangements. The species most suitable for the garden grow upright to about 2 m (6 ft) and are like little trees in form. The leaves are long and slender, rather like those of a willow in shape but they do not hang down. The

China Flower *(Adenandra uniflora)* is effective in the garden or in a pot.

flowers are small and carried in dainty clusters in spring. The large seed-pods which follow are decorative on the plant and in vases. The sap of the plant is milky. Stems cut for arrangements should be stood in boiling water, which just covers the bottom end of the stem, for a minute or two. This ensures that they will last well.

CULTURE: Wild cotton grows very easily and quickly from seed and, under good conditions, tends to seed itself too freely. The seeds inside the large pods are beautifully arranged, each with its own silky parachute to help in its distribution. It is recommended for dry gardens and for those who grow flowers for the cut-flower trade. Sow the seed in late summer or spring.

A. fruticosa WILD COTTON, MELKBOS

Has seed pods which are oval rather than round and densely covered with soft prickles. They appear in summer and early autumn. This species is common in warm parts of the Transvaal and in the Karoo.

A. physocarpa WILD COTTON, WILDE KAPOK

In summer this species bears huge, globular seed pods like large balls, coloured pale green flushed with maroon. It is the more decorative of the two species.

The scented flowers of South African Gorse (Aspalathus) brighten the garden in spring.

Wild Cotton *(Asclepias physocarpa)* bears decorative seed pods.

ASPALATHUS — South African Gorse

DISTRIBUTION: There are as many as 150 species growing in South Africa, the most decorative being found in the eastern and south-western Cape.

DESCRIPTION: The name is derived from the Greek *aspalathos* meaning "scented shrub". Some of them have flowers with a delightful scent like vanilla, whilst others give off a faint indefinable scent. They are woody shrubs generally with tiny leaves rather like those of an erica. The flowers are pea-shaped and carried in loose clusters or along the ends of stems. They are usually yellow, sometimes turning brown as they age so that yellow and brown flowers appear on the same plant. Unfortunately they have not as yet all been named. Many of them are fine plants for the rock - garden and they look splendid planted amongst other shrubs in a shrub border.

CULTURE: They grow from seed or cuttings and thrive in poor soil and in sandy soil near the coast. They are not hardy to severe frost but tolerate moderate frost. Seed should be sown in spring and cuttings made from spring to the end of summer.

A. capensis
(*A. sarcodes*)
Grows to 2 m (6 ft) and more with very small leaves only about 10 mm long. The yellow pea-shaped flowers about 2-3 cm in length are carried at the ends of the stems and are very showy from spring to early summer.

A. macrantha
Reaches a height of 2 m (6 ft) and has tiny leaves like an erica and yellow pea-shaped flowers in late spring.

A. spinosa
Is a tufty little shrub to about 60 cm (2 ft) with erica-like leaves and masses of canary-yellow flowers.

ASTER FILIFOLIUS — Wild Aster

(Diplopappus filifolius)
[New name is *Felicia filifolia*]
DISTRIBUTION: Common on dry hillsides from the south-western Cape eastwards into the Little Karoo and north into the Orange Free State and the Transvaal.

DESCRIPTION: Is a showy little shrub in late winter and early spring when it is in full flower. It grows to about 30-45 cm (1-1½ ft) in height and spreads across as much or more. The leaves are slender and rather like those of an erica, and the flowers are like daisies with very slender mauve petals and a yellow centre. They measure 2-3 cm across. It is a decorative plant for the flower border or the front of a shrub border. It also makes a good show in a rock-garden or when grown in containers.

CULTURE: It can be grown from seed or cuttings. The plant stands quite considerable frost, but it should be watered during winter. It grows well in poor soil and in coastal gardens.

ATHANASIA COULTER BUSH, KOULTERBOS

DISTRIBUTION: Decorative species are to be found growing along the roadsides and lower mountain slopes in the south-western Cape, in Natal and the Transkei.

DESCRIPTION: These are quick-growing plants which vary in size according to species. They have decorative foliage and colourful flowers in mid-spring to summer and are splendid plants for large gardens where there is enough space for them to spread. In small gardens they should be trimmed back to keep them within bounds. This trimming will also help to prevent them from becoming woody at the base, which often happens if they are allowed to grow on without any pruning after their flowering period is over. They are good plants to use as a stop-gap in a new garden where they will provide colour until the slow-growing plants are mature. They make an effective show lining a large avenue or drive. Some species could be made more use of also in roadside planting.

CULTURE: They grow readily and quickly from seeds or cuttings and do well even in poor soil. In fact they can become a troublesome weed and they should not be planted in gardens where they are likely to invade the lands. Species from the south-western Cape should be watered in winter whilst those native to the summer-rainfall region need water in the summer months. If cut back by frost they usually grow up again quickly.

A. acerosa

This species from the eastern Cape and Natal sends up many stems from the ground, densely clothed with slender finely-divided leaves. In warmer parts of the country it might over-run the garden but in gardens where winters are fairly cold, it is unlikely to grow very large.

Wild Aster *(Aster filifolius)* grows well in poor soil.

A. crithmifolia KLAAS LOUW BOS

In mid-spring this species makes the roadsides in the south-western Cape gay with its golden-yellow flowers carried in flat clusters. The narrow little leaves scarcely 1 mm broad are divided at the ends into three little "fingers". The plants grow to a height of about 1.5 m (5 ft) with a spread of nearly as much. The common name is said to be derived from the fact that a farmer with this name allowed this species to grow unchecked on his lands, with the result that it spread and became a nuisance on neighbouring farms!

A. parviflora COULTER BUSH, KOULTERBOS

Grows to about 3 m (10 ft) in height with a spread across 3-4 m (10-13 ft), but can be kept smaller by pruning after the flowering period. The base of old plants tends to become bare and leggy if the plant is not cut back. It bears large flat clusters of flowers of a rich golden-yellow from October to November. The clusters measure 7-10 cm across and are composed of minute flowers closely packed together. The leaves which have an aromatic scent are delicately cut and of a soft shade of green, making the plant decorative even when it is not in flower.

AULAX AULAX

DISTRIBUTION: Occurs in the south-western Cape.

DESCRIPTION: Aulax is a member of the protea family and worth growing in gardens where conditions suit it. As in the leucadendrons, the male and female flowers are carried on different plants. The male plants bear catkins of flowers at the ends of the stems whilst the female ones produce cones. The leaves differ according to species.

CULTURE: Plant them in acid soil, and water them well during winter and spring. It is advisable to make holes 30-60 cm deep and across, and to incorporate an abundance of acid compost in the holes. If watered well they endure fairly severe frost.

A. pinifolia AULAX

In manner of growth and height this species is similar to the above, but its leaves are quite different. They are rather like those of a pine, but softer and slightly curved, and they are carried all along the stem. In spring the male plant bears flowers similar in form to the above-named species, but they are not as showy. The female plant is smaller and has small cones which are yellow to lime-green shaded with deep coral-red.

A. umbellata AULAX
(A. cneorifolia)

in this species the male plant is particularly showy. It grows erect to 2 m (6 ft) or more and has narrow and somewhat leathery leaves 5-6 cm long, which are broader at the apex than at the base. They tend to hug the stem rather than stand out at an angle. The little flowers of sulphur-yellow are carried in conical spikes arranged in a whorl at the top of the stem. A plant in full flower is most decorative. The flowering time is late spring and early summer.

BARLERIA OBTUSA BARLERIA

DISTRIBUTION: About fifty species of barleria occur in Southern Africa half of which grow in the Transvaal. This species is to be found in the warmer parts of the Transvaal, in Natal and the coastal regions of the eastern Cape.

DESCRIPTION: It is a decorative evergreen shrub growing to 1 m (3 ft), but under optimum conditions it will send its soft, twiggy growth much higher than this and, if it is planted amongst other shrubs it will frequently grow up and over them. The leaves are oval, often curved over at the end and covered with fine hairs. The flowers are

Coulter Bush *(Athanasia parviflora)* is a quick-growing shrub for the large garden.

The male flowers of Aulax *(Aulax umbellata)* are most decorative.

usually mauve but there is a pink form too. The plant flowers in profusion in autumn and looks particularly decorative when grown next to a shrub which has yellow flowers at the same time. The flowers consist of a short tube opening to a face with five segments of uneven shape measuring 2-3 cm across. Barleria does well in seaside gardens and at fairly high elevations.

CULTURE: This is a quick-growing shrub particularly in warm regions. It will stand moderate frost but should be given some protection where winters are severe. Even when cut to the ground by frost it generally grows quickly again and will flower by the following autumn. In warm gardens it should be cut back each year to keep it within bounds. This is best done immediately after flowering or in early spring. It grows easily from cuttings taken during the warmer months of the year. Water the plants in summer.

BAUHINIA GALPINII PRIDE OF DE KAAP

DISTRIBUTION: Grows wild in the eastern Transvaal. The common name is derived from the fact that it was originally found in abundance in the Kaap Valley near Barberton.

DESCRIPTION: This is a vigorous shrub which can be grown as a specimen alone, in a mixed border of shrubs, or as the boundary planting in a large garden. It is a useful plant for large gardens because of its spreading habit of growth. When grown in a small garden it is advisable to trim the plant back once a year to keep it within bounds. It grows quickly to a height of 2 m (6 ft). The leaves have rounded lobes arranged so that they look like a pair of wings. The flowers are carried in showy clusters at the ends of stems. Each flower is made up of spatulate petals, i.e. narrow at the base and wide and rounded at the ends. They are apricot-red to brick-red in colour and the plant is very attractive in summer and early autumn when it bears most of its flowers.

CULTURE: This robust shrub thrives in warm places where there is an abundance of sunshine. It stands mild frost but severe frost is apt to cut the plant back. In gardens which experience more than five degrees of frost it is advisable to plant it where it has some protection. This is a quick-growing shrub which is not particular as to soil and which will endure long periods without water.

BERKHEYA BARBATA PRICKLY
(B. ilicifolia) SUNFLOWER, DORING GOUSBLOM

DISTRIBUTION: Can be found on lower mountain slopes and near the sea in the south-western and north-western Cape.

DESCRIPTION: This is a rounded shrub with a height and spread of 60-90 cm (2-3 ft). The leaves are not very large and have indented margins with sharp prickles. Both stems and the undersides of leaves are felted grey. The seed-heads are

Barleria *(Barleria obtusa)* showing close-up of flowers.

Barleria becomes wreathed in flowers in late autumn.

Brunia *(Brunia laevis)* is effective in arrangements.

Pride of de Kaap *(Bauhinia galpinii)* is a colourful, quick-growing shrub.

also full of prickles. The flowers which appear in spring, are like daisies, 6-8 cm across and of a bright yellow colour. *B. decurrens* is a showy species which grows wild in the eastern Cape. Several other species occur in other parts of the country.

CULTURE: These are useful plants to provide a background of colour in gardens where the soil is poor and where growing conditions are not good as, once established, they stand drought and fairly severe cold. If they become too large for their allotted space cut the plants back a little after their flowering period. They grow easily from cuttings or seeds.

BERZELIA LANUGINOSA KOLKOL

DISTRIBUTION: Its natural habitat is the south-western Cape.

DESCRIPTION: Is an upright-growing shrub to 2 m (6 ft) with tiny, soft, needle-like leaves closely arranged along the stems. It has minute flowers in little round balls the size of a large pea. They are green at first becoming ivory-white and fluffy with red appendages when mature. This is a decorative shrub for any part of the garden and useful in providing long-lasting material for arrangements. The flowering time is from late winter or early spring to summer.

CULTURE: Kolkol likes an acid soil and plenty of moisture, thriving in marshy conditions. When grown in the summer-rainfall region it should be watered well and regularly throughout the autumn and winter. Plant it in holes to which plenty of acid compost has been added. It tolerates quite severe frost, but in hot inland districts it should be planted where it is partially shaded during the heat of the day.

BRUNIA BRUNIA, STOMPIE

DISTRIBUTION: Can be found in the south-western Cape but is no longer as widespread as it once was in this area.

DESCRIPTION: The plants grow to 1-2 m (3-6 ft) or more, and have leaves rather like those of an erica. The leaves are neatly arranged along the length of the stem, embracing it and pointing upwards rather than outwards. The flowers are carried in round, or almost round heads, which are in clusters at the ends of stems.

CULTURE: Plant them in soil in which there is plenty of acid compost and some peat, and keep

◁ The flowerheads of Kolkol *(Berzelia lanuginosa)* are unusual in form.

The Prickly Sunflower *(Berkheya barbata)* does well in coastal gardens.

them well watered from autumn to spring. They grow naturally in ground which is often marshy in winter. They stand fairly severe frost when once established.

B. laevis

This plant reaches a height of 60-90 cm (2-3 ft) and has tiny leaves which hug the stem tightly near the top. The flowerheads are round and about 1.5 cm across. They are unusual rather than pretty and the plant makes an interesting specimen in a damp part of the garden.

B. nodiflora STOMPIE

Grows to 1 m (3 ft) with very small leaves hugging the stem and flowers which look like little pincushions. They are ivory in colour, faintly scented and open in winter to early spring.

B. stokoei BRUNIA

Grows to about 2 m (6 ft) and is erect in habit. In early spring it bears its attractive heads of flowers which look like ornamental buttons, flat on top and rounded beneath. These are greyish to ivory-white.

BURCHELLIA BUBALINA WILD POMEGRANATE, WILDEGRANAAT

DISTRIBUTION: Occurs from Swellendam eastwards up into Natal and the warmer regions of the Transvaal.

DESCRIPTION: This evergreen shrub, which grows to 2-3 m (6-10 ft) and more, is ornamental throughout the year for it has attractive dark green, glossy foliage. The flowers, carried in a little whorl at the ends of stems, are tubular ending in five turned back petals. They are 2-3 cm in length, and of a bright coral to tomato-red colour which shows up beautifully against the green of the leaves. The buds are clothed with silky hairs which make them glisten in the sunlight. The flowering time is mid-spring to early summer.

CULTURE: The wild pomegranate is a slow-growing plant which does best in gardens near the coast or in regions where winters are not severe and where there is an abundance of rain. Plant it in a hole to which plenty of compost has been added. Away from the coast it should be planted where it is shaded by a tree or building during the hottest hours of the day. In cold gardens it should be planted in a protected position as it is tender to frost.

CADABA APHYLLA — DESERT BROOM, SWARTSTOOM

DISTRIBUTION: Can be seen in arid country in the Karoo.

DESCRIPTION: This is a straggly shrub useful for difficult gardens because of its tolerance of frost and aridity. It reaches a height of 1 m (3 ft) when standing alone but if it is between other plants it grows up and through them to flower at the top of them. It has hardly any leaves except on the new growth. In late spring it bears a profusion of crimson flowers.

CULTURE: It does not need good soil and tolerates a wide range of growing conditions. It grows well in alkaline soil. To keep the plant neat in appearance trim it back lightly after it has finished flowering. It can be grown from seed or root sections.

CALPURNIA — CALPURNIA

DISTRIBUTION: Different species of calpurnia are fairly widely distributed in the eastern Cape and Natal.

DESCRIPTION: These plants vary in height from 1-3 m (3-10 ft). The divided leaves are rather graceful in appearance. The flowers are pea-shaped, yellow in colour and carried in attractive sprays.

CULTURE: Calpurnias generally grow readily without much attention other than regular watering when young. They endure moderate frost. The plants can be grown from seed or cuttings.

C. floribunda reaches a height of 2 m (6 ft) and needs fairly good soil and some shade in hot gardens. In late summer and early autumn it bears its drooping panicles of golden-yellow flowers.

C. intrusa is 1 m (3 ft) or more in height and spread with yellow flowers.

C. villosa bears large pendant trusses of yellow flowers on plants 3 m (10 ft) high. It looks rather like a laburnum.

CARISSA — AMATUNGULU, NUM-NUM, NATAL PLUM

DISTRIBUTION: Are scattered in many parts of the eastern Cape and in warmer parts of Natal and the Transvaal.

DESCRIPTION: These are plants of decorative value because their evergreen glossy, dark leaves are pleasing throughout the year. The leaves are oval in shape, darker on the upper surface than the underside. The flowers are white and sweetly scented, rather like a jasmine, both in appearance and in scent. The fruit is bright scarlet, and both flowers and fruit show up well against the leaves. They are good plants to use anywhere in the garden and they make a first-class impenetrable hedge. The strong sharp thorns allow neither animal nor man to penetrate a hedge of well-grown plants. The flowering time is late spring and summer.

CULTURE: These plants thrive near the coast or in

◁ Wild Pomegranate *(Burchellia bubalina)* is a pretty shrub for warm areas with a good rainfall.

gardens inland where winters are mild. They can be grown in gardens which have moderate frost but the rate of growth is much slower in areas where winters are cold.

C. bispinosa NUM-NUM

Grows to 1-2 m (3-6 ft) and has sturdy stems armed with rigid spines. The leaves are oval and deep green. The small, white flowers are carried in dainty little clusters. It flowers in spring and later small, scarlet fruits which are edible brighten the plant.

C. macrocarpa AMATUNGULU

(*C. grandiflora*)

Grows to 3 m (10 ft) but can be kept trimmed back to smaller size. It has long strong spines. The leaves are dark green, shiny and leathery and the white flowers with five slender oval petals are very sweetly scented. These are followed by scarlet fruit the size of a small plum, from which a delicious jelly can be made. The plant has a milky juice. It grows particularly well in warm, coastal gardens where there is an abundance of moisture in the air.

CHRYSANTHEMOIDES MONILIFERA BOETABESSIE

DISTRIBUTION: Is to be found in many parts of the south-western Cape—in sand along the coast, on plains and on lower mountain slopes. Related subspecies occur in many parts of Southern Africa.

DESCRIPTION: This is an exuberant shrub, eager to grow and quick in growth. If left untrimmed it may grow to a height of 2.5 m (8 ft) with a spread of almost as much. For this reason it is advisable to trim the plant back annually after its flowering period is over. The leaves are a pleasing shade of green, wider at the apex than the base, with uneven and widely spaced notches. In September it becomes covered with little clusters of bright yellow daisies which measure 2 cm across, and when these fade the shrub bears masses of berries, carried in tight clusters of three to five together. The common name of boetabessie is derived from "boetie", meaning brother, referring to the close arrangement of the berries. These are relished by birds and children.

CULTURE: This is a shrub for the large garden or park rather than the small garden as it is inclined to take up too much space. It seems to thrive in any soil and it stands drought and moderate frost.

CLEMATIS BRACHIATA TRAVELLER'S JOY, OLD MAN'S BEARD, KLIMOP

DISTRIBUTION: This plant is to be found in many parts of Southern Africa, except the south-western Cape.

DESCRIPTION: It is a climbing plant which trains itself over any support near it, whether it be a stake or another plant. It supports itself by twining the leaf-stems sharply around anything handy. The leaves are divided and sharply indented. The ivory-white flowers have a more pronounced scent in the evening than during the day. They are carried in great profusion, making a wonderful sight in late summer and autumn. Each flower is composed of four pointed sepals with a crown of stamens ornamenting the centre. The seed-pods which follow the flowers grow into feathery round heads which are ornamental too. One African tribe is said to prepare an infusion from the leaves for the treatment of worms.

CULTURE: It tolerates moderate frost but not long periods without water, and should be watered fairly regularly during spring and summer to promote good growth and flowering. Add compost or old manure to the holes in which it is planted.

CLEMATOPSIS SCABIOSIFOLIA BUSH

(*C. stanleyi*) CLEMATIS, SHOCK-HEADED PETER, PLUIMBOSSIE

DISTRIBUTION: Grows wild in the central Transvaal and in Rhodesia.

Amatungulu *(Carissa macrocarpa)* grows well near the coast and in warm gardens inland. ▷

DESCRIPTION: This is a decorative, shrubby plant growing to 1 m (3 ft). It has leaves which are much divided, giving it a rather feathery appearance. The flowers, which measure about 5 cm across, or more, are bowl-shaped and drooping. Each flower has four broad sepals which look like petals. They are pale pink or mauve with showy yellow stamens crowning the centre. The seed heads which follow the flowers have long silky hairs and make a pretty show, too. Some tribes rub the flowers in their hands and then inhale this to clear a cold in the head.

CULTURE: The plant, which flowers in spring and summer, dies down in winter and needs little water at this time, but it should be kept watered regularly from when the new shoots emerge in winter and until autumn. Although it grows well in any kind of soil it undoubtedly will perform better if given some attention. It can be grown from seed sown in autumn or spring or from root divisions.

COLEONEMA PULCHRUM CONFETTI BUSH, PINK COLEONEMA

DISTRIBUTION: Grows wild in the south-western Cape as far east as Port Elizabeth.

DESCRIPTION: This is a charming plant for the garden, ornamental at any time of the year, particularly if it is clipped to shape as a round ball, or square-cut. It can be used as a hedge, as a clipped specimen in a tub on a patio or unclipped, amongst other shrubs in a shrubbery. It grows to 1.5 m (4-5 ft) in height and spread, and has tiny, needle-like leaves growing all along the stems. They give off a rather pleasant pungent smell when crushed. The flowers are very small but carried in such profusion that the bush is a cloud of pink when it is in full flower, which is late winter or early spring. There is another pretty species, *C. album*, which has white flowers.

CULTURE: This decorative shrub grows very well in coastal gardens and it also does well in highveld gardens if sheltered from frost. In hot dry gardens it should be planted where it is shaded by a tree or wall for the hottest part of the day and watered regularly during autumn and winter. It grows quickly from seed or cuttings.

COMBRETUM BURNING BUSH, RUSSET BUSH, WILLOW, HICCUP-NUT

DISTRIBUTION: There are many species of combretum in different parts of South Africa and Rhodesia. Some are shrubs and some are small trees.

DESCRIPTION: See under the species given below.

CULTURE: Once established the species described will stand moderate to severe frost as well as fairly long periods without water. Generally they are quick-growing in a warm climate, particularly if planted in holes to which plenty of compost has been added, and if watered regularly.

C. bracteosum HICCUP-NUT

This plant, which grows wild in the eastern Cape, the Transkei and Natal, is of scrambling habit. It will grow to 3 m (10 ft) over other plants and produce its flowers above them. The leaves are oval and pointed and about 7 cm long. The five-petalled flowers are brick-red and carried in rounded whorls at the ends of the stems. The buds have a velvety brown and green sheen. The flowering time is spring, and later the plant bears winged fruits which are used by some Bantu tribes for the relief of hiccups. This is a useful plant for coastal gardens.

C. hereroense RUSSET BUSH,
(*C. transvaalense*) WILLOW, KIERIEKLAPPER

A large shrub or small tree which is common in the lowveld and bushveld of the Transvaal and is to be found also in Natal and Rhodesia. It grows to about 3-4 m (10-12 ft) and has small oval leaves about 2-3 cm long with a dark green surface. The underside is pale green with clearly defined veins. The flowers are inconspicuous but they develop

◁ Boetabessie *(Chrysanthemoides monilifera)* is a quick-growing plant for large gardens or for roadside planting.

into pretty four-winged fruits coloured green and russet, which festoon the plant in autumn and produce quite a show. It stands dry conditions and fairly sharp frost.

C. microphyllum BURNING BUSH, FLAME CREEPER

A large, deciduous scrambling shrub which can be found in the northern and eastern Transvaal and Rhodesia. In spring it bears tiny crimson flowers in such quantities that it looks as though the stems are covered in flames. They have minute bright crimson stamens which light up the whole plant. The leaves vary in size being much larger at the bottom than at the top of the plant. This plant needs a hot situation for best development.

CROTALARIA BIRD FLOWER, CAPE LABURNUM

DISTRIBUTION: Different species are widely distributed all over Southern Africa although most of the showy ones come from the eastern and northern part of the country.

DESCRIPTION: They vary in height from low-growing perennials to large shrubs. The leaves are arranged in threes, and the flowers are like those of a pea in shape, usually yellow or mauve in colour and arranged in clusters or spikes.

CULTURE: They grow readily and quickly in any reasonably good soil but are tender to sharp frost and should be protected when grown in gardens which have severe frost.

C. agatiflora BIRD FLOWER

This is a pretty species which grows wild in East Africa. It grows rapidly to a height of 2-3 m (6-10 ft) and has long sprays of flowers of a charming shade of greenish-yellow. It grows well in any soil or situation but is apt to be cut back by sharp frost. It is seldom killed, however, and usually grows up rapidly again in spring and flowers by summer.

C. capensis CAPE LABURNUM

Although it grows well in the south-western Cape and is now common it is thought to be native to the eastern Cape, Natal, the eastern Transvaal and Rhodesia. In cultivation it grows to 2-3 m (6-10 ft), but in nature it seldom reaches more than a height of 1 m. The trifoliate pale green leaves give the plant a graceful appearance and the attractive racemes of yellow flowers add to its value in the garden. It flowers mostly in spring but may have some flowers at other times of the year too.

C. mucronata
(C. striata)
Reaches a height of 1.5 m (5 ft) and has flowers of sulphur-yellow.

C. purpurea
Grows to about 2 m (6 ft) and produces attractive stems of deep pinkish-mauve flowers in late winter and early spring.

CYCLOPIA GENISTOIDES BUSH TEA, HONEY TEA, HEUNINGTEE

DISTRIBUTION: Occurs in the south-western Cape usually on hillsides and mountain slopes.
DESCRIPTION: This is a decorative plant which grows to about 1.5 m (5 ft) and is erect in habit with branches from low down. The leaves are soft and needle-like and arranged in threes. The pea-shaped flowers are of golden-yellow and carried in clusters at the ends of the stems. Each flower is only about 2-3 cm long but they make a wonderful show in August and September when the plant is in full flower. The foliage and young flowers were used by early settlers to produce a tea with an aroma of honey.
CULTURE: This cyclopia does not tolerate severe frost and when grown in the summer-rainfall area it should be watered regularly during autumn and winter. Plant it in ground to which compost or manure has been added. It grows from seed or cuttings.

Confetti Bush *(Coleonema pulchrum)* makes a charming picture in spring. ▷

White Confetti Bush *(Coleonema album)*. (See page 76.)

DISSOTIS CANESCENS — Wild Lasiandra

(D. incana)

Distribution: Is found in the eastern Cape, the Transkei, Natal, Swaziland, the north-eastern Transvaal and Rhodesia.

Description: It is a shrubby plant growing to about 1 m (3 ft) and sometimes more in height and spread. The leaves are oval and pointed with clear ribs and a soft velvety texture very like the lasiandra commonly grown, which is an exotic. The five-petalled flowers are 3-4 cm across and of a delightful shade of pinkish-mauve with showy magenta stamens in the centre. It makes a most attractive show when it flowers in summer. Another species worth trying is *D. princeps*, which grows to 2 m (6 ft) and has violet flowers. They flower in late summer and autumn.

Culture: These plants like growing where their roots can reach down to moisture. They should be planted at the edge of a pond or next to a tap which drips, or be watered well. They are not hardy to severe frost but frequently grow up again quickly after being cut down by frost. Prepare holes 45 cm deep and wide and fill them with compost and peat which helps to retain moisture around the roots.

Burning Bush *(Combretum microphyllum)*. (See page 77.)

Crotalaria makes a bright show in spring. (See page 77.)

DOMBEYA — DOMBEYA

DISTRIBUTION: Can be seen in the eastern Cape and the warmer parts of Natal and the Transvaal.

DESCRIPTION: Dombeyas vary in size from shrubs to trees. They are quick-growing plants some of which have attractive leaves and clusters of flowers which are showy. These unfortunately sometimes spoil the appearance of the plant when they fade, for the faded flowers persist instead of falling. To keep the plant attractive cut off the flowers as they fade.

CULTURE: They grow naturally in regions which have regular rain and which are fairly warm, and they do not do well in gardens where severe frosts are experienced. Given protection when young they will however stand fairly cold winters when mature. They grow readily from cuttings.

D. burgessiae — PINK-AND-WHITE DOMBEYA

Is a quick-growing species from the warm parts of Natal and Zululand. It reaches a height of 2-3 m (6-10 ft) in two or three years, and in late summer bears rounded heads of white flowers with pink marks and a faint scent. The large leaves, measuring up to 20 cm in length, are heart-shaped and soft in texture. It does best in a temperate and humid climate.

D. tiliacea — HEART-LEAF DOMBEYA

(*D. dregeana*)

Can be seen in the eastern Cape near the coast, through the Transkei and in Natal. It has large, soft heart-shaped leaves about 10 cm or more in length with crenated margins and white fragrant flowers in loose clusters. Each flower has wide petals arranged in bowl formation. It is a quick-growing plant.

DOVYALIS CAFFRA — KEI APPLE, DINGAAN'S APRICOT

DISTRIBUTION: Grows wild in the eastern Cape, the Transkei and Natal.

DESCRIPTION: This small tree or large shrub is a useful one for making a hedge or screen in areas where climatic conditions are difficult. It grows to 3 m (10 ft) in height but can grow much higher, and has pale green oval leaves. The flowers are not attractive but the orange-yellow fruit rather like an apricot makes quite a show in late summer. They are reputed to make a good jelly. The plant is armed with spines 3-5 cm long which add to its value as a hedge plant. It could be made more use of in large farm gardens to make the boundary of

Wild Lasiandra *(Dissotis princeps)* is a quick-growing shrub with pretty flowers.

Bird Flower *(Crotalaria agatiflora)*. (See page 77.) ▷

Bush Tea *(Cyclopia genistoides)* is a pretty sight when in full flower. (See page 77.)

the garden, or to enclose an orchard or a camp for stock.

CULTURE: Although it is native to the warm coastal area of the country the Kei apple will stand quite severe frost and it also endures long periods of drought when once it is established. It is a good plant, therefore, to use in gardens where low rainfall and hot days followed by cold nights make gardening difficult. It grows easily from cuttings.

DUVERNOIA ADHATODIOIDES PISTOL BUSH
(Adhatoda duvernoia)

DISTRIBUTION: It occurs in the eastern Cape, the Transkei and Natal, usually on the edges of forests.

DESCRIPTION: The pistol bush grows to a height of 2.5 m (8 ft) and is an evergreen with pretty, dark-green, oval leaves which show off to advantage the heads of white to mauve flowers. Each flower consists of a tube about 2 cm long opening to two lips which are streaked with maroon or crimson. They have a faint scent. It is a fine shrub for the border and makes a good hedge too. The common name is derived from the fact that the seeds are exploded with a sharp report when ripe. The flowering time is from March to May.

CULTURE: This shrub is tender to sharp frost but has been grown successfully in some highveld gardens where it has protection. It does well in full sun near the coast but inland it should be planted where trees will shade it for part of the day. Provide it with good soil in which there is plenty of compost, and water it liberally throughout the year to promote quick growth.

ENCEPHALARTOS CYCAD, BREAD TREE, BROODBOOM

DISTRIBUTION: Can be found near the coast in the eastern Cape and Natal and inland, in the bushveld and lowveld of the Transvaal.

DESCRIPTION: These fascinating and decorative plants generally known as cycads are a link with the past inasmuch as they represent the descendants of plants found fossilized in sedimentary rock deposited 150 million years ago. They have not changed in form and look today as they must have looked when dinosaurs roamed the countryside. It is difficult to estimate the age of existing specimens. Those few which stand 8 m tall could be as much as 500 years old, and the rootstock of species which produce suckers from the base may have attained a much greater age.

They look rather like palms or large tree ferns. Some are stemless with huge leaves arching up in whorls from ground level, and some have thick stems which grow to 8 m (25 ft) and carry their whorls of leaves at the top of the stem. The leaves live for several years, and when they drop there is a well-defined scar on the stem where they were attached. Cycads carry their male and female flowers on separate plants, but it is not possible to determine the sex unless the cones are present. Some of the female cones are like huge pineapples in form and golden-yellow or orange in colour. The seeds and their outer covering are said to be poisonous. A cycad makes a fine accent plant in the garden where growing conditions suit it.

The name of Bread Tree is said to have been derived from reports made by early travellers, who intimated that Hottentots and some Bantu tribes used the starchy pith in the trunks for making a kind of bread.

CULTURE: Cycads are very long-lived, but they are slow to germinate from seed and their subsequent growth is very slow also. It is a pity that nurseries did not propagate these plants many years ago so that plants would now be available. They need high humidity, warmth and shade to encourage germination. It is thought that in nature seeds germinate best when not covered by soil. It is possible also to propagate them by removing basal suckers which may or may not have roots. If watered regularly these will eventually produce roots. Some growers have also raised them by planting a small piece of stem attached to the base of a leaf.

E. altensteinii BREAD TREE, BROODBOOM
A tall-growing species from the eastern Cape which reaches a height of about 5 m (16 ft). It is a graceful and dramatic plant.

E. ferox
Sends its handsome leaves up from the ground or near ground level. Each leaf grows to 1-2 m in length and spreads out from the centre of the plant. This species occurs in the coastal bush in Mozambique.

E. horridus PRICKLY CYCAD
Produces long arching grey-green leaves at ground level. They are about 1.5 m (5 ft) long and twisted along their length. The leaflets are tough with forked ends and have sharp spines.

E. natalensis NATAL CYCAD
This is a tall species which grows to about 5 m (16 ft) and has a crown of arching palm-like leaves at the top. It is a very handsome plant suitable for gardens where winters are mild.

E. transvenosus MODJADJI OR MUJAJI CYCAD
Grows to 9 m (30 ft) and is to be found growing wild near Duiwelskloof in the Transvaal. Some specimens of this one can be seen in The Wilds, Johannesburg. The female cones of this species may weigh as much as thirty kilograms. Bantu people living in the area in which these plants grow regard them as sacred.

The cone of a Cycad. *(Encephalartos ferox).*

The Natal Cycad *(Encephalartos natalensis).*

E. villosus KAFFIRBREAD TREE, WILD DATE

This species which grows wild in the coastal forests near East London does best in partial shade. Its leaves, which rise in whorls from the ground, develop a tuberous underground stem. They measure up to 2 m in length. The female cone of this one gives off an unpleasant odour. The plant can be propagated from young shoots from the underground stems planted as cuttings. Seeds of this and other cycads are used by some African witch doctors for necklaces and as a charm to ensure long life.

ERICA ERICA, HEATH

DISTRIBUTION: There are approximately 600 species of erica in South Africa, and of these about 580 occur in the south-western Cape. In the admirable book *Ericas in Southern Africa* by Baker and Oliver, it is indicated that in the Caledon district alone there are some 220 species, whilst along the strip from the coast to the mountains parallel to the coast, extending from Mossel Bay to Humansdorp, there are 112 species; and, in the Cape Peninsula, there are 103 known species. It will be noticed that there are few growing wild in other parts of South Africa. There are also very few species in the whole of the rest of the world.

Most of the ericas of Southern Africa occur on coastal plains and on mountain slopes in areas where the rainfall is regular from autumn through winter and until spring, and moisture at this time of the year appears to be one of the important factors in promoting their growth. A few of them grow where the precipitation may be more than 2000 mm a year and where there is snow in winter, but many species grow in regions which become extremely hot and dry during the summer months. With most garden plants there is a general rule that they should be watered more when coming into flower than at other times of the year, but this does not apply to all of the ericas, for some of them flower in their natural haunts during the dry, summer months.

DESCRIPTION: Having such a large number of native species of erica of incomparable beauty it is surprising that so few gardeners in Southern Africa have tried to grow them. Some of our ericas were being grown in England nearly 200 years ago, generally in glass-houses, and Australian and New Zealand gardeners have been growing some of them for many years, too. In these two countries they are also grown in fields for sale to florists, whilst here in their homeland, it is still rare to find growers producing these lovely flowers for the cut-flower trade.

Ericas vary tremendously in size, from those a few centimetres high to large shrubs which may grow to 2-3 m or more in height. Their leaves are small, short, and sometimes needle-like, but the flowers show great differences in appearance. They may be like little bells only about 3-5 mm long, or they may be flask-shaped, urn-shaped or tubular, measuring up to 3 cm or more in length. The texture of the flowers varies a good deal as well. They may be transparent, glossy, waxy or papery. The range of colours of the flowers of ericas is considerable, embracing many shades of pink to deep crimson and scarlet, white, mauve, cyclamen, purple, pale to deep yellow, orange and green.

CULTURE: Perhaps one of the reasons why ericas have not been more widely grown is because they have very specific needs. First of all the growing period of most ericas is winter and early spring and they must be kept watered at this time even although they may not flower until summer. Secondly most of them need acid soil to promote growth. Some species grow best in very acid soil whilst others do well in soil which is only slightly acid and a few grow in alkaline soil. Gardeners who wish to grow a number of ericas should have the pH of the soil tested. Should the pH be around 5 it is safe to plant ericas, but if it is higher than 6 only a few species will thrive. The term pH defines the degree of acidity or alkalinity of soils measured on a scale ranging from 0 to 14. Soils with a pH of 7 are neutral whilst those with a pH above 7 are alkaline and those with a pH below 7 are acid. The Department of Agriculture, Pretoria, can advise gardeners where to send soil for testing. There are also soil-testing kits available from firms which stock garden sundries. Generally it can be assumed that where hydrangeas bear flowers which are distinctly blue the soil is likely to be acid, whereas if their flowers turn pink the soil is likely to be alkaline.

The texture of the soil does not appear to be of such great importance. Sometimes ericas are found growing luxuriantly in poor sandy soil; sometimes in gravel, sometimes in good loam, and sometimes in clay.

Where ericas are to be grown in regions where the soil is not acid, peat and acid compost, such as that derived from wattle or pine leaves, should be forked into the soil. The easiest way to do this is to make a trench 60 cm (2 ft) deep and wide, and to fill it with acid compost and peat together with some of the soil previously removed from the trench. Should the water also be alkaline it is advisable to take further action to prevent the pH of the soil being changed too drastically by the water. Regular applications of soil sulphur or aluminium sulphate (alum) will help to keep the soil acid. A teaspoonful sprinkled around each plant once a month is effective.

To have success with ericas in dry, inland districts they should be watered regularly, particularly from late autumn to early spring when many species are making most growth. When grown in gardens where severe frosts are experienced, set them out so that they are shaded by trees or a building until about 10 a.m. This will prevent the early morning sun from scorching the frosted leaves. Some ericas grow naturally on mountains where they are under snow fairly often and these species would be more able to endure severe frost, but not drought.

Ericas may be propagated from cuttings made from the tips of the stems. The cuttings should be approximately 4-8 cm in length. They can also be grown from seed. Gardeners will generally find it more rewarding to order their plants from nurseries which specialize in propagating ericas. Whether growing them from seed or cuttings it is wise to keep the containers in a shady place until growth is certain. Sow the seed in late summer or spring and make cuttings in spring or summer. There are nurseries which propagate many of the ericas described below and they supply plants which are already a year or two old. When transplanting the plants obtained from a nursery try to get them out of the containers without disturbing the roots. If it is a plastic container cut down along the side. If they are in tins, water the soil in the tin thoroughly, then loosen the soil around the edges with a knife or tap the side of the tin against something hard, turning the tin so that the soil is loosened all round. This generally makes it possible to get the soil out in one piece.

Where to plant Ericas. These lovely plants can be grown in any part of the garden, the smaller species combined with annuals or perennials and the larger ones planted at the back of the flower border or in front of shrubs. But, since one has to

The porcelain flowers of the Bottle Heath *(Erica ampullacea)* are very attractive.

Erica aristata is pretty in the garden and in arrangements.

Bridal Heath *(Erica bauera)*. A charming species for the garden or for growing in large containers.

Berry Heath *(Erica baccans)* has small, but lovely flowers.

watch the acidity of the soil, it is in most cases advisable to plant them together in one place so that it is easier to ensure that they receive the attention they need. Many of the smaller ericas make splendid pot plants and can be grown in a patio or on a balcony or windowledge.

E. ampullacea BOTTLE-HEATH, SISSIE-HEATH

This species which grows in the countryside near Caledon reaches a height of about 60 cm (2 ft) and is rather straggly in habit. Plants should be cut back after flowering to keep them more shapely. The beautiful flowers compensate however for the somewhat unattractive nature of growth of the plant. They are carried on stems emerging from the tops of the branches and are flask-shaped opening to four wide segments. The flowers measure 2-4 cm in length and look as though they have been fashioned from the most exquisite, delicate and lustrous porcelain. They are white and palest blush-pink marked with rose. It flowers from late winter to summer.

E. ardens WAX HEATH

This species from the mountains near Swellendam and Riversdale grows to 45 cm (1½ ft) in height and is bushy in form. It bears masses of tiny urn-shaped flowers in spring. The flowers, which are pink, measure only about 8 mm in length and the sepals, which are also coloured, almost cover the corolla. There is a white form of this which is greatly prized for bridal bouquets. It is known as the Riversdale Bridal Heath. This species stands frost well.

E. aristata

This charming species found on mountain slopes near Stanford in the Caledon district, is one of the most attractive of the ericas and well deserves a place in the garden. The plant grows to 60 cm (2 ft) and has unusual leaves. They are oval and saw-toothed along the margins. The flowers are carried in fours at the ends of stems. They are flask-shaped and 2-3 cm long, broader at the bottom than at the top with an open face composed of four rounded segments. The flowers glisten and are beautifully coloured white and purplish-rose. It flowers from August to October and in March to May.

E. baccans BERRY HEATH

This species, which occurs in the Worcester-Clanwilliam area reaches a height of about 1 m

Bridal Heath *(Erica bauera)* — showing detail of flowers ▷ and leaves.

The Lantern Heath *(Erica blenna)* with its richly-coloured flowers is a fine plant for pots or the rock-garden.

(3 ft) and has most decorative globular flowers of clear pink flushed with rose. They are carried in fours and in such masses that one hardly sees the leaves which lie flat against the stem. When it flowers in late winter and early spring it makes a lovely show.

E. bauera ALBERTINIA HEATH,
(*E. bowieana*) BRIDAL HEATH
A species with great charm whose natural habitat is the Albertinia-Riversdale region. It has been grown successfully in gardens in many parts of South Africa and abroad. The plant is upright in growth to a height of 1 m (3 ft) or more, and has the typical small leaves, but they often stand straight out from the stem rather than along it. The lovely translucent, waxy flowers of palest pink, rose, cyclamen or white appear close together hanging down near the ends of the stems. It flowers most in spring but has some flowers on and off throughout the year. It is delightful in small arrangements.

E. bergiana FAIRY ERICA
Occurs in marshy places on mountains near Stellenbosch to Worcester and Ceres. This is a rangy species growing to 1.5 m (5 ft). The flowers appear to be carried in fours at the ends of the stems and droop prettily. Each little flower is only about 3 mm long. They are urn-shaped and the ends of the flowers point straight out instead of being recurved as in many other species. The sepals of this species are completely reflexed. The flowers are of lovely shades of pink to rose. The flowering time is summer.

E. blandfordia BLANDFORD'S HEATH
This is an erect species which grows near Worcester and Tulbagh often at high altitudes. It reaches a height of 1 m (3 ft) and has fairly long spikes of glistening yellow flowers at the ends of the stems. Each flower is only about 5-7 mm long and is a short squat tube with a narrow throat. It flowers in late spring and stands dry growing conditions well.

E. blenna LANTERN HEATH, RIVERSDALE HEATH, ORANGE HEATH
This species can be found on mountain slopes near Swellendam. It is a most decorative one for growing in pots or the garden. The plant reaches a height of about 60 cm (2 ft) and has flowers measuring up to 18 mm in length. They are

broadly urn-shaped, waxy in texture and of glowing orange with green tips. The flowers are often grouped in threes near the tips of the branches.

E. blenna var. **grandiflora** is another outstanding species with larger flowers up to 3 cm in length. They both flower in late winter and early spring.

E. bodkinii
This is a rare species found occasionally in the Caledon-Bredasdorp area, in places which are marshy in winter. It grows to about 60 cm (2 ft) and is erect in habit of growth. The neat slender little leaves hug the stems and the flowers are carried close together at the ends of stems. They are ivory in colour with the sepals forming a skirt half over the corolla. A very pale pink form has also been found. The flowers are hairy and beautifully formed and about 12 mm in length.

E. borboniaefolia FLOUNCED HEATH
This is a species which is found fairly high up on mountain slopes in the Caledon district. The plant grows to a height of about 60 cm (2 ft) and has minute spear-shaped leaves. The flowers are carried in clusters at the ends of stems coming off the main branches. Each one is flask-shaped, much broader at the bottom than at the top and about 10 mm long. They are rose in colour. The bracts and the sepals are the same colour as the flower, and the sepals partially enclose the corolla, making the flower look like a series of flounces. The flowering time is summer.

E. caffra HEDGE HEATH
Its natural habitat is along streams from the Cape Peninsula to Natal. This is a tall species which grows to a height of 3-4 m (10-13 ft). The stems and leaves are grey-green and the white flowers are urn-shaped and about 8 mm long and pleasantly scented. The bush tends to become untidy as the faded flowers do not drop off. It is hardy to frost and the individual stems of flowers are useful for making posies. The flowering time is late winter and early spring.

E. campanularis YELLOW BELL HEATH
This is a fine species which seems to favour wet places along the sides of streams or marshes. It is a slender erect plant to 45 cm (1½ ft). The leaves are carried erect along the stem. The flowers are bell-shaped and of a clear yellow colour, each flower measuring 8 mm in length. They are carried on short stems close together on the upper ends of the branches, forming rather loose

The Grahamstown Heath *(Erica chamissonis)*.

Blandford's Heath *(Erica blandfordia)*.

Red Hairy Erica *(Erica cerinthoides).*

spikes of colour. The flowering time is late winter and early spring.

E. cerinthoides RED HAIRY ERICA

This is the most widespread in its distribution, being found from the Cape Peninsula east through to Natal and from there to Lesotho, Swaziland, the Transvaal and Rhodesia. It varies considerably in growth according to habitat, being generally bushy and not more than 30-90 cm (1-3 ft) in height in areas where veld fires burn it to the ground, and rather lanky and straggly to 2 m (6 ft) in other areas. This popular erica was introduced into cultivation in England as long ago as 1774, and with its showy flowers it is well worth a place in gardens in South Africa. The flowers are carried in drooping clusters at the ends of the stems. Each flower is tubular in form, 2-4 cm in length and covered with fine, downy hairs. They are of a rich rose to crimson colour and very attractive. There is also a white form of this one.

This form of *Erica cerinthoides* can be found growing wild in the Eastern Cape and north, into the Transvaal.

The plants are hardy to frost. It flowers from late winter to late spring but carries some flowers on and off throughout the year. It is advisable to cut this species back hard occasionally to keep it compact in shape.

E. chamissonis GRAHAMSTOWN HEATH
A species, limited in distribution to the countryside near Grahamstown in the eastern Cape, grows to about 60 cm (2 ft) and more, with upright branches and leaves which are often hairy and arranged in threes. It bears spikes of cup-shaped, cyclamen flowers about 8 mm across. The flowering time is spring but some flowers appear at other times of the year too. This is a good species for gardens where soil is inclined to be alkaline.

E. chloroloma GREEN-EDGED HEATH
A tall species which occurs between Mossel Bay and Port Elizabeth often in limestone areas. It grows to a height of 2.5 m (8 ft) and has upright branches. The tubular crimson flowers are about 16 mm long. The word *chloro* meaning green and *loma* meaning fringe, describe the green tip of the flower. This is a colourful shrub which should be tried in areas where the soil is not acid. The main flowering time is late autumn to early spring.

E. coccinea SMALL TASSEL HEATH
(*E. petiveri*)
This species, widely distributed north of Cape Town to Clanwilliam and eastwards to Knysna, is an erect-growing plant to a height of 1 m (3 ft) or more. It has soft leaves in tufts along the stem. The flowers are arranged in drooping clusters along the ends of the branches. A characteristic of the flower is the way the long brown stamens show well below the corolla and the sepals hug the corolla. Each flower is 15-25 mm in length and there are different colour forms, varying from crimson to yellow. *E. coccinea* var. *melastoma* has showy yellow bracts and sepals and a purplish-brown corolla.

E. corifolia
Is one of the commonest species found in the south-western Cape where it is often found in poor sandy soils on plains and mountain slopes and sometimes in moist places. It grows to about 45 cm ($1\frac{1}{2}$ ft) and bears little clusters of pale pink to rose flowers. The flowers vary considerably in size according to habitat. The urn-shaped flower is almost entirely enclosed within the cup-shaped calyx. Both are tinged with pink. The flowering time is on and off from spring through summer.

E. curviflora WATER HEATH, WATERBOS
This species occurs from the Cape Peninsula north to Clanwilliam and eastwards to Grahamstown. It grows to 30-90 cm (1-3 ft) in damp places and along the edges of watercourses. The older branches are almost bare of leaves but on the young branches the leaves are closely crowded. The flower consists of a slender slightly curved tube of salmon, rose or orange colour. They measure about 2-3 cm (1 in) in length and are sometimes hairy, but not always. The flowering time is summer but they may appear at other seasons, too.

E. curvirostris SCENTED BELL HEATH
Found in the Cape Peninsula and towards Worcester and Caledon. This species is not showy but it is worth growing because of the scent, which few ericas have. It grows to about 60 cm (2 ft) and in late summer and autumn has small clusters of white or rose bell-shaped flowers. Each flower measures 3-5 mm in length. It does not appear to be as particular as to soil as many other species.

E. daphniflora THE DAPHNE ERICA
Occurs on mountains and plains, from Clanwilliam south to Ceres, Tulbagh and Worcester, and eastwards towards Swellendam. It grows erect to 60 cm (2 ft). The flowers appear in rather tight clusters making spikes along the ends of the stems. They vary considerably in colour and include ivory, pale pink, deep rose and yellow. Each flower is less than 12 mm long and flask-shaped with a face made up of curved back petals rather like those of the daphne. In full flower this species makes a gay show. The main flowering time is spring and summer.

E. decora STICKY ROSE HEATH
This is a very pretty erica found in the Cape Peninsula. It would make a handsome addition to any garden. The plant grows to a height of 60 cm (2 ft) and is loosely branched with leaves sparsely arranged along the stem. The bell-shaped flowers are of a charming shade of mauve to cyclamen. Its flowering time is autumn to spring, and it should be watered regularly to induce good flowering.

E. deliciosa PORT ELIZABETH HEATH
Grows from Mossel Bay to Port Elizabeth in

The Daphne Heath *(Erica daphniflora)* adds beauty to the spring garden.

Sticky Rose Heath *(Erica decora)* produces its dainty flowers during winter and early spring.

damp or dry places in fields and mountain slopes. The plants generally reach a height of 1 m (3 ft) but may grow even taller. The urn-shaped flowers with projecting stamens are white or pink in colour and tightly clustered along the upper parts of the branches. Each little flower measures only about 4 mm in length but they are carried in such profusion as to make a splendid show when in full flower between June and December.

E. densifolia KNYSNA HEATH
Is one of the most attractive heaths growing wild in the Knysna area. It grows to about 1 m (3 ft) and bears tight clusters of leaves all along the branches. The flowers spray out in rounded clusters towards the ends of the stems. Each flower is a curved tube about 2.5 cm long and usually crimson, attractively edged at the mouth with greenish-yellow. The flower is often covered with downy hair. It flowers in September to March.

E. eugenea
This charming erica which occurs in the Cedarberg and Cold Bokkeveld area, where winters are often severe, grows to 30 cm (1 ft) and has dark green slender leaves which stand up at an angle to the stem. The corolla is urn-shaped and about 12 mm long and coloured white to palest pink. It is partially enclosed by the long pink sepals. The bracts are large and coloured pink to rose, tipped with green. This is a dainty and most attractive species.

E. fascicularis TIGERHOEK HEATH
Can be seen in fields and on mountain slopes up to about 1000 m near Paarl and Stellenbosch and east to Swellendam. It grows to about 1.25 m (4 ft) but may be much taller than this. It does not branch much but the stems end in a wheel of slender flowers, tubular in form, flushed with shades of pink and tipped with green. They are glossy and sticky to the touch. The flowering time is mainly autumn and winter. This is a spectacular species for garden show.

E. fastigiata FOUR SISTERS HEATH
This species, which grows naturally in damp places near Paarl, Worcester and Caledon, is an erect plant reaching a height of about 45 cm ($1\frac{1}{2}$ ft). The flowers are of unusual form somewhat different to most of the other ericas. Each flower consists of a tube broader at the base than the apex, which opens up to a starry face with four widely spaced lobes. These four lobes and the fact that the flowers are carried in fours account for its common name. The flowers are ivory to palest pink in colour flushed with rose and the faces of the little flowers have a distinct darker rose or green ring around the centre. The flowering time is spring. This is a very attractive species for the garden and for arrangements.

Erica eugenea is one of the prettiest of the ericas. ▷

The Large Orange Heath *(Erica grandiflora)* makes a bold display in summer.

Petticoat Heath *(Erica glauca* var. *elegans)* is decorative in the garden and in pots or window-boxes.

E. formosa
The word *formosa* means beautiful and this heath, which grows between Mossel Bay and Humansdorp, is certainly well named. It reaches a height of about 60 cm (2 ft) with spreading or erect branches. The flowers are carried in threes, each one being like a tiny globe. They measure only 4 mm across and are waxy in texture and white in colour. This is a dainty flower, ideal for corsages and small arrangements. It stands a good deal of frost. The flowering time is winter and spring.

E. glandulosa STICKY-LEAFED HEATH
Occurs from Mossel Bay to Port Elizabeth. It grows to about 1.5 m (5 ft) but should be kept cut back to smaller size as, left to develop on its own, it becomes rather straggly. This is not one of the more attractive of the ericas but it grows so readily that it is worth trying in gardens where conditions may not be suitable for other species. It is hardy to frost. The plant is sticky and the flower is wax-like in appearance and made up of a slender slightly curved tube about 2-3 cm long, shaded from apricot to a salmon-pink. The flowering time is mainly May-September.

E. glauca CUP-AND-SAUCER HEATH
This shrubby species found on the mountain slopes near Ceres, grows to about 1 m (3 ft). It bears several flowers in whorls on gracefully arching stems. The flower is urn-shaped with a slender neck and is about 10 mm long. They are unique in that the colourful effect is derived not only from the corolla of the flower, which is plum to rose, but also from the sepals and bracts which are large and of the same shades. The term *glauca* describes the grey-blue colour of the leaves. It flowers in winter and spring.

E. glauca var. **elegans** PETTICOAT HEATH
Occurs in the Bain's Kloof mountains near Wellington and on mountain slopes at Tulbagh, Piketberg and Elgin. This one is even more attractive than the species described above. In this variety the corolla is palest pink tinted green or lilac and partially hidden by the most attractive calyx which may be cyclamen-pink, alabaster or fawn. The colourful bracts make an additional flounce behind the sepals, and the whole flower head droops elegantly and is most attractive. A single plant growing in a pot is very decorative. In the garden it will grow to 1 m (3 ft) or more.

Cup and Saucer Heath *(Erica glauca* var. *glauca)* has charming flowers of unusual colouring.

E. grandiflora LARGE ORANGE HEATH
Its natural habitat is on mountain slopes from Stellenbosch and Paarl to Ceres, and east towards Robertson and Ladismith. The plant is robust and well worth a place in the large garden. It grows to about 1.5 m (5 ft) and is bushy in habit. The leaves are like those of the pine but much shorter and carried close along the branches. The glossy, sticky flowers which are arranged in spikes on the upper parts of the stems are very attractive. Each flower is tubular with lobes opening out, about 2-3 cm long and of a glowing orange colour. In newly opened flowers the prominent anthers add to their attractiveness. The var. *exsurgens* has flowers of sulphur-yellow. They endure hot, dry growing conditions well.

E. grisbrookii
Is a vigorous plant reaching a height of about 60 cm (2 ft). In winter it bears flowers about 12 mm in length. Each flower is composed of a corolla, white or ivory in colour, with the sepals forming an overskirt of the same colour. A well-grown plant makes a handsome show. It tolerates fairly dry conditions.

E. haematosiphon
This erica is rather straggly in appearance and

the flowers are variable in colour. At their best they make a fine show. The flowers which are about 16 mm in length are like slender tubes with an enlarged end. They hang down from the plant, attached to the main stems by short, slender stalks. The colour is rose or crimson. This plant will stand considerable cold and fairly long periods of drought. The flowering time is summer and autumn. The plant reaches a height of 30 cm or more.

E. hibbertia
This heath, which grows on the slopes of the Franschhoek Pass, was introduced to England early in the nineteenth century. The plant grows to 60 cm (2 ft) and has whorls of flowers near, or at the ends of stems densely clothed with short stubby leaves which point upwards all along the stem. The flowers are sticky and glossy, each one a slender curved tube 2-3 cm long, coloured rose to crimson at the base and tipped with yellow and green at the top—a lovely combination of colours. The flowering time is spring.

E. holosericea SMALL FLOUNCED HEATH
This species found in the Caledon district is a fine plant to grow in pots because the beauty of the flowers can be the more fully appreciated when seen at eye level. In the garden the plants grow to about 1 m (3 ft). The flowers are rather drooping in habit of growth. Each one is composed of a corolla which is somewhat bell-shaped and only about 8 mm long. The coloured sepals almost entirely cover the corolla so that only a flounce of the cyclamen corolla shows up below the overskirt of alabaster to pale pink sepals. When the corolla fades the pink sepals remain attractive for quite a long time. The flowering time of this enchanting species is spring.

E. imbricata
This species, found in many parts of the southwestern Cape, is variable in colour and form depending on where it is growing. It is a bushy plant to about 1 m (3 ft) and becomes covered with minute dainty flowers which are usually white, but may be pink, brown or russet in colour. They measure only 8 mm in length and have prominent dark anthers which show below the flowers. The flowering time is late winter and spring.

E. inflata
Grows from Ceres across the Bokkeveld to the Cedarberg. The plant reaches a height of 1 m (3 ft) and is much branched. It bears rounded heads of pale to dark cerise flowers at the ends of stems. Each flower is only about 8 mm long and broadly urn-shaped. They appear in September to May, the hot dry months of the year in their natural habitat.

E. infundibuliformis FUNNEL HEATH
Occurs in damp places on the mountains near Paarl, Worcester and Caledon. The long specific name of this one means nothing more than "shaped like a funnel". The long, slender tubular part of the funnel is pink to rose in colour, and the starry face is pink or white. The flowering time is summer.

E. irregularis GANSBAAI HEATH
This pretty species which can be found in limestone areas between Hermanus and Gansbaai should be tried in gardens where the soil is alkaline. It is rather like *E. baccans* in appearance but the flowers are arranged differently on the stems. The plant grows to about 1.5 m (5 ft) and the flowers emerge from the stem forming a spike towards the ends of the stems. The little stems carrying the flowers have silky hairs and are coloured pale pink and droop down to hold the charming little flowers of the same colour. The sepals of clear pink are snugly set around the lower half of the urn-shaped corolla. The flowering time is late winter and early spring.

E. junonia
Grows on mountain slopes in the districts of Ceres and Clanwilliam. This plant has not yet been brought down from its mountain stronghold and tamed for gardens, but perhaps in the near future some enterprising nurseryman will make the effort to propagate it. It certainly deserves to be cultivated for it is a striking plant. I cannot add to the description given in *Ericas in Southern Africa* by Baker and Oliver . . . "No other erica can match *E. junonia* for the magnificence of its show of flowers. The species not only possesses the largest flowers in the genus but also has a shape and colour that are unique".

The specific name is derived from Juno, the queen of the gods in Greek mythology. In its natural habitat it grows to only 30 cm (1 ft) and bears flowers in clusters at the ends of stems.

The flowers of the Small Flounced Heath *(Erica holosericea)* are enchanting in form and colour. ▷

Erica lanipes is a dainty flower for pots or window-boxes as well as in the garden.

Each flower is flask-shaped, much broader at the bottom than at the neck, and each one ends in a starry face composed of 4 oval, pointed lobes. The whole flower measures about 4 cm in length and is rose-coloured and waxy. It flowers in summer.

E. lanipes

The distribution of this erica seems to be limited to higher mountain slopes in the Caledon district. It is an erect shrub to 1.25 m (4 ft) with leaves neatly pressed around the stems. The flowers which are of a charming and delicate shade of mauve are carried along the upper parts of stems. Each flower is urn-shaped with reflexed segments. This is a graceful erica for any part of the garden.

E. lanuginosa — BEAKED ERICA

This species found near Caledon has an unusual flower and should be grown in pots as well as in the garden. It has flowers measuring about 2 cm in length. The corolla is broad at the base with lobes tightly closed at the tip giving the flower the appearance of the beak of a bird. The sepals half enclose the corolla and are light green tinged with rose at the tips. It grows to 60 cm (2 ft) and flowers in winter.

E. lateralis — GLOBE HEATH

The globe heath is fairly widespread on mountain slopes in the south-western Cape, from Paarl to Tulbagh and Ceres, and east towards Caledon. The plant grows to 45 cm ($1\frac{1}{2}$ ft) and is as much across. It bears little urn-shaped flowers 8 mm long, arranged in a spike along the top part of the stem. The flowers are rose-pink in colour and hang down gracefully. The flowering time is summer.

E. longifolia — LONG-LEAFED HEATH

Its natural habitat is in the country between Paarl and Bredasdorp. The specific name indicates that the leaves are long and slender, but in fact they are sometimes rather short. The plant grows to 1 m (3 ft) in height and has long, tubular flowers with four lobes forming a starry face. They are arranged in several whorls near the top of the stem. Each flower is 2-3 cm long and hairy. They vary in colour from white to pink, rose, yellow, orange, or purple. It flowers mainly in summer but bears some flowers at other seasons too.

E. lutea

This species which grows in sandy soils on mountains near Caledon, Paarl and Worcester reaches a height of 1 m (3 ft) and bears its leaves in pairs opposite each other and tightly pressed against the stem. The flowers are usually carried in twos or fours at the ends of stems. Each flower is an elongated urn 8 mm long and yellow or white in colour. In summer when it is in full flower it makes a fine show.

E. macowanii

Occurs naturally in the Caledon district. This erica grows in an erect fashion to between 1 and 2 metres and bears its flowers rather closely clustered along the ends of stems, making a loose spike. Each flower is tubular and about 2 cm in length, with slightly reflexed lobes. They are beautifully coloured red at the base and yellow at the tip, or they may be plain yellow.

Plant Red Signal Heath *(Erica mammosa)* for a striking show in the rock-garden or shrub border.

Erica macowanii is effective in the garden and in arrangements.

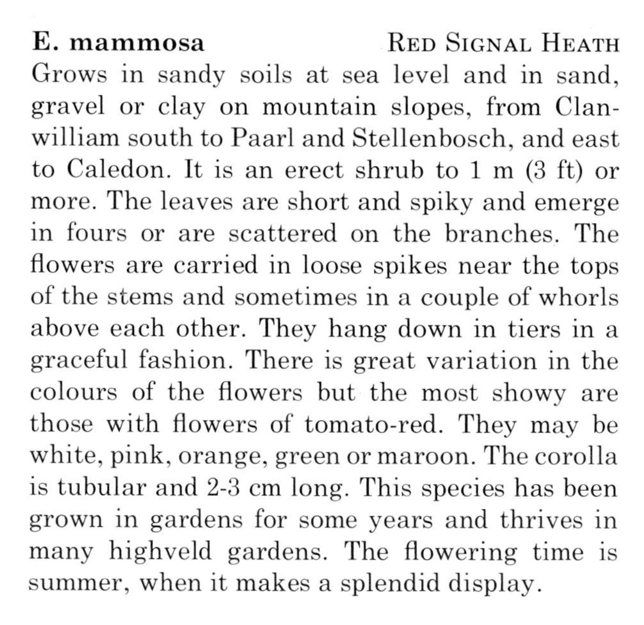

E. mammosa RED SIGNAL HEATH

Grows in sandy soils at sea level and in sand, gravel or clay on mountain slopes, from Clanwilliam south to Paarl and Stellenbosch, and east to Caledon. It is an erect shrub to 1 m (3 ft) or more. The leaves are short and spiky and emerge in fours or are scattered on the branches. The flowers are carried in loose spikes near the tops of the stems and sometimes in a couple of whorls above each other. They hang down in tiers in a graceful fashion. There is great variation in the colours of the flowers but the most showy are those with flowers of tomato-red. They may be white, pink, orange, green or maroon. The corolla is tubular and 2-3 cm long. This species has been grown in gardens for some years and thrives in many highveld gardens. The flowering time is summer, when it makes a splendid display.

E. massoni HOUHOEK HEATH

This plant from the Hottentots Holland mountains is very showy when it is in full flower. It grows to 45 cm ($1\frac{1}{2}$ ft) and has oval, hairy leaves which hug the stem and curve outwards. The flowers are sticky and glossy and carried in whorls of 5-10 flowers at the ends of the stems. The flowers are 2-3 cm long and orange or crimson in colour, beautifully edged with pale and deep green at the tip. The flowering time is summer.

E. nana DWARF HEATH

It is a low-growing species which scrambles along stones and soil on the cliffs near Hangklip. The flowers are carried in threes or fours together. They are about 2-3 cm long, tubular in form with a constricted neck and recurved lobes. They are greenish-yellow when they first open and turn bright yellow as they mature, making a fine splash of colour. This species is an excellent plant for the rock-garden. The flowering time is spring.

E. oatesii

Its natural habitat is the summer-rainfall regions,

where it is to be found in rocky mountainous country in Natal, the Orange Free State, Lesotho, Swaziland and the Transvaal. The plant grows to 1.25 m (4 ft) and has spreading branches. The leaves are generally in threes or fours and about 8 mm long. The flowers are carried in little groups on stems coming off the main stem. They are short, tubular and pink to scarlet in colour. It flowers from March to August.

E. ovina WOOLLY ERICA

This erica from mountain slopes near Caledon and Worcester reaches a height of 1 m (3 ft) and has short tufts of leaves. The flowers form a cylindrical spike near the top of the stem. They are bell-shaped and covered with hairs which give them a woolly appearance. The corolla is 6-9 mm long and is white. There is an even prettier one *E. ovina* var. *purpurea* which has pink flowers. They flower mostly in spring.

E. parilis

This species grows on mountain slopes in the districts of Clanwilliam, Ceres, Tulbagh, Worcester and Paarl where the summer is dry and hot. It grows to 60 cm (2 ft) and is erect in habit. The leaves are spear-shaped, very small and hug the stem. The flowers are broadly urn-shaped with reflexed lobes at the mouth. They appear close together to form a spike at the ends of the stems. Each flower is only 4-6 mm long and bright yellow in colour. This is a showy little flower which appears in summer despite the intense heat and aridity of its habitat at this time of the year.

E. passerina

This is an unusual erica which grows on mountain slopes east of George. It reaches a height of 60 cm, is erect in growth and bears very pretty bell-shaped flowers about 8 mm across. They are covered with a silky, woolly fuzz and are of a charming shade of pinkish-mauve.

E. patersonia MEALIE HEATH

Can be found growing along plains and lower slopes between Hangklip and Kleinmond, always fairly near the coast. It sends up a few erect stems to 1 m (3 ft) or more. The leaves are densely arranged around the stems and are slender and slightly curved. The flowers make cylindrical spikes along the top section of the stem. They are tubular, about 1.5 cm long and tend to curve slightly. They are of a glowing shade of canary-

◁ *Erica passerina* has dainty, decorative flowers.

yellow. This is a decorative plant and useful since it provides flowers in winter. It is hardy to frost, but when grown in gardens subject to severe frost it should be given some protection.

E. perspicua PRINCE OF WALES HEATH
Its natural home is the Caledon district where it is usually seen growing in areas which are fairly damp. This is a striking heath and particularly fine for arrangements. It grows erect to about 1 m (3 ft) or more, and has plumes of flowers (reminiscent of the feathers in the crest of the Prince of Wales). Each flower is slightly curved and tubular. They are 2-3 cm long, translucent in texture and shaded pink to cyclamen at the base and white at the top. It flowers in late winter and early spring.

E. peziza VELVET BELL HEATH, KAPOKKIE
Occurs near the coast from Cape Town east towards Caledon and inland to Montague. It grows to 1.5 m (5 ft) and is rather bushy in form. The flowers are minute little velvety white bells carried in great profusion. They have a delightful scent of honey. This is an excellent plant for the garden. It stands quite considerable frost. The flowering time is spring.

E. pillansii PILLANS' HEATH
Can be found in the Caledon district and near Kleinmond in marshy places and along the sides of streams. This species grows to 1.25 m (4 ft). The leaves are arranged in tufts and are incurved. It bears showy spikes of coral-red tubular flowers, each flower being about 18 mm long. The flowering time is autumn and winter.

E. pinea
This is a robust species which grows to 1.5 m (5 ft) and is upright in habit. It bears showy spikes of flowers in spring, and on and off during other months of the year, too. Each flower is tubular in form and 2-3 cm in length and of an attractive shade of canary yellow, although occasionally those of paler shades may be found. It should be cut back in late spring to keep it neat in form.

E. plukeneti TASSEL HEATH
This species, which grows from Cape Town to Mossel Bay and north to Namaqualand, is not as showy a heath as many other species, but, as it is not as particular as to soil as some of the others, it is recommended for gardens where the soil is alkaline. It grows to 60 cm (2 ft) and has a cluster of carmine flowers near the ends of the stems. Each flower is tubular and 18 mm in length.

Prince of Wales Heath *(Erica perspicua)* is a striking plant for the garden and in arrangements.

Tassel Heath *(Erica plukeneti)* has unusual flowers.

Velvet Bell Heath *(Erica peziza)* bears a cloud of minute flowers with the scent of honey.

Royal Heath *(Erica regia)* makes a dazzling show.

Erica pinea is a good species for the large garden.

Royal Heath *(Erica regia)* showing colour variation.

Green Heath *(Erica sessiliflora)* produces flowers of subtle colouring.

Varieties of this one are to be found in different colours. They all have stamens which project far out from the mouth of the flower. It flowers in spring.

E. porteri
Occurs near Betty's Bay and is named after Mr Harold Porter, the founder of the Porter Reserve there. It grows to 1 m (3 ft) and has slender leaves which stand out horizontally from the stem. The shining tubular flowers are carried near the ends of the stems. Each one is deep crimson or rose shaded to white at the mouth of the flower. They are 2-3 cm long and very attractive. The flowering time is winter.

E. quadrangularis PINK SHOWER HEATH
This species can be found from the Cape Peninsula north to Clanwilliam and east to Knysna. It reaches a height of 45 cm (1½ ft) and has slender little leaves all along the stems. The tiny flowers which are pale pink, rose or white are generally carried in fours and make a loose spray along the upper sections of the stems. Each flower is broad and squat, almost bell-shaped and 3-4 mm long. Although very small they are carried in such profusion that the plant is most colourful in spring when it is in full flower.

E. regia ROYAL HEATH, ELIM HEATH
This lovely heath which grows wild near Elim in the Bredasdorp district is fairly well known, and grown in gardens in various parts of the country. It is a spectacular plant when in flower and should be grown more extensively both in gardens and as a pot or tub plant. It grows to 60-90 cm (2-3 ft) and bears beautiful glossy flowers of crimson to scarlet. Each one is tubular with five lobes turned back at the mouth. The most beautiful form of this is known as *E. regia* var. *variegata*. Its flowers are luminous and shaded from gleaming white to red at the tips and sometimes to green. The flowers are about 18 mm long and hang down in a whorl from the ends of the stems. The main flowering time is late winter and early spring.

E. sessiliflora GREEN HEATH
Can be seen from the Cape Peninsula north to Piketberg and Tulbagh and east to Humansdorp, frequently in places which are marshy in winter, but it is no longer found in abundance. The plant grows erect to 1.25 m (4 ft) or more and bears spikes of flowers made up of whorls of ice-green, translucent flowers arranged above each other. The flowers are tubular, about 2-3 cm long and slightly curved downwards. After the corolla drops the sepals enlarge and remain green for a while finally turning brick-red. This is a most decorative flower which is at its best from autumn to spring. The plants should be trimmed back each year or two to keep them from becoming straggly.

E. shannonea STAR-FACED HEATH
This rare species sometimes found near Hermanus is a compact plant growing to 45 cm (1½ ft). Its leaves are rather spear-shaped and often hairy. The flowers are carried in rounded heads at the ends of the stems. Each one is a tube inflated at the base and contracted at the neck, ending in a starry face of four lobes. They are 3-4 cm long and white, sometimes suffused with rose. The flowering time is summer. It is a showy plant well worth cultivating.

E. sparrmanni
Although the plant in nature tends to be straggly it can be kept trimmed in the garden to improve its shape. It grows to 45 cm (1½ ft) and bears ovoid flowers of an unusual shade of lime-yellow. They are covered with rather bristly hairs and arranged in little clusters which hang down around the plant. The flowering time is spring and summer.

E. taxifolia DOUBLE PINK HEATH
Grows on rocky or sandy hill slopes near Paarl, Tulbagh, Worcester and Caledon. It reaches a height of 60 cm (2 ft) and is a neat plant in growth with leaves rather like those of a yew, standing out from the stem at an angle. The flowers are carried in loose sprays near the tops of the stems and although the flowers are small (8 mm long) they are very showy. Each flower is broadly urn-shaped and pink to rose in colour. The sepals are also pink and large, half enclosing the corolla, making the flower look double. It flowers in summer and makes a very fine show in a pot or tub as well as in the garden. It stands drought well.

E. thunbergii MALAY HEATH
This is a spectacular little plant from the mountains near Ceres and Clanwilliam and grows to about 30 cm (1 ft). It bears little bowl-shaped flowers on drooping stems. The flowers are bright orange whilst the sepals and bracts are lime-

yellow, making an exciting and gay combination of colour. This species is worth trying in gardens and pots. It flowers from September to November.

E. tumida MOUNTAIN HEATH

Grows on the higher slopes of mountains from Worcester to the Cedarberg. This is a robust heath which is 1 m (3 ft) tall and across. It bears striking heads of flowers of crimson to coral-red. The flowers are often arranged in fours at the upper ends of the stem, each one being broadly tubular in form, and about 2-3 cm long, and hairy. It should be more widely grown. Its flowering time is late spring and summer—a dry hot period in its natural habitat.

E. ventricosa WAX HEATH, FRANSCHHOEK HEATH

The wax heath, which occurs on mountains near Paarl and Stellenbosch, is fairly tall, reaching a height of 1m (3 ft), but it can be kept cut back to smaller size. The leaves, arranged in fours, stand out almost at right-angles to the stem. The flowers are exceptionally pretty. They are carried in rounded clusters. Each flower is bottle-shaped and about 18 mm in length, broad at the base and narrow at the neck, opening to a starry face of 4 lobes. The flowers are delicately shaded pink and white, and have a most attractive sheen. This species was grown in England during the last

The Malay Heath *(Erica thunbergii)* bears flowers of arresting colours.

Double Pink Heath *(Erica taxifolia).* This pretty species adds colour to the garden in summer.

The Wide-mouthed Heath *(Erica vestita)* stands considerable frost.

The flowers of Wax Heath *(Erica ventricosa)* add charm to the garden in late spring.

century and it certainly deserves to be more widely grown in gardens and in large pots or tubs. The flowering time is late spring and early summer.

E. versicolor
Can be found in many parts of the Cape, from the Peninsula east to George. It is rather straggly in habit of growth and should be trimmed back occasionally to keep it neat and tidy. It grows to 1-2 m (3-6 ft) and bears its flowers in groups. Each flower is a slender tube with a definite curve and charmingly coloured bright red with a green mouth. They measure 2-3 cm in length and are sometimes slightly sticky. This is a robust species for gardens large and small.

E. vestita WIDE-MOUTHED HEATH
This species grows on mountains near Caledon and Worcester. It reaches a height of 1 m (3 ft) and has stems densely covered with soft needle-like leaves which give the plant a graceful feathery appearance. The flowers are tubular in form with a wide mouth and are grouped near the tops of the stems. Each flower is 2-3 cm in length and crimson, white or pink in colour. The plant has been grown successfully on the highveld and is hardy to frost. It flowers in spring and summer.

E. viridiflora GREEN HEATH
Can be seen on mountain slopes from Mossel Bay to George and Knysna. Although the plant is rather untidy the green flowers are so attractive that it deserves a place in the garden. Each flower is a slender, glistening, sticky tube of vivid green fading to brown as it ages. They measure 2-3 cm in length and are carried in drooping little clusters at the ends of stems. The plant grows to 1 m (3 ft). It flowers in summer. In order to keep it neat it is advisable to trim it back every year or two.

E. walkeria WALKER'S HEATH, SWELLENDAM HEATH
Occurs on mountain slopes near Wellington and Tulbagh and towards Swellendam. The flowers are 12 mm long and flask-shaped, opening out to a starry face of four lobes. They vary in colour from white to palest pink and rose. The flowers have a luminous quality and look as though they are fashioned from porcelain. They make a lovely show in winter and early spring when they are most abundant. The plant is neat in appearance and grows to 60 cm (2 ft). The leaves are soft and sparsely arranged on the main stem and more closely together on the flower stems, which come off the main stem at intervals.

Erica versicolor is a large plant suitable for the shrub border.

E. woodii WOOD'S HEATH
This is one of the twelve species which come from the Transvaal. It occurs also along the mountains of Lesotho, Swaziland, Natal and the eastern Cape. It grows to 60 cm (2 ft) and is spreading in habit. It carries long sprays of minute flowers delicately arranged on the stem. Each flower is only about 2 mm in length and of pale pink or white. They are like globular bells in form. It flowers in summer and autumn.

ERIOCEPHALUS AFRICANUS WILD ROSEMARY, KAPOKBOSSIE
DISTRIBUTION: There are many species of eriocephalus, but the most decorative are the two mentioned from the south-western Cape.

DESCRIPTION: This shrub grows to about 1 m (3 ft) or more and spreads across as much. It has tiny, grey, needle-like leaves which give off a pleasant

The Dwarf Kaffirboom *(Erythrina humeana)* has flowers which produce a ravishing show of colour.

Wild Rosemary *(Eriocephalus africanus)* will show up well in the flower or shrub border.

aromatic scent when crushed. The small flowers are white or palest pink. They appear in winter, but the plant is more attractive in spring when it goes to seed, as the seed heads look like fluffy little pieces of cottonwool, and the effect of the whole plant is grey and white. It makes a fine foil to flowers of blue or yellow. *E. punctulatus* is very similar in appearance. The fluffy heads last long in arrangements.

Culture: It grows readily from seeds or cuttings and is able to stand long periods without water and a good deal of frost. The plant should be watered in winter and early spring.

EROEDA IMBRICATA

Distribution: Occurs in the Cape Peninsula, Clanwilliam and neighbouring areas and east towards Grahamstown.

Description: This is an angular little shrub which grows to 30-60 cm (1-2 ft) in height and spread. The leaves are variable in shape and size. Sometimes they are oval and pointed and sometimes they are almost as broad as they are long. Often they are larger near the flower than lower down on the plant, and generally they are recurved. In September the plant becomes covered with bright, orange-yellow daisies which measure 4 cm across.

CULTURE: This is a useful shrub for seaside gardens. It will stand moderate frost and drought.

ERYTHRINA TAMBOEKIE THORN

DISTRIBUTION: Can be found from the eastern Cape through Natal into the Transvaal and Rhodesia.

DESCRIPTION: Although the tree erythrina (Kaffirboom) is well known and popular, few gardeners are aware of the beauty of the shrubby types described below which are well worth a place in the garden. They have showy flowers of tomato-red to scarlet.

CULTURE: These plants grow best in districts where there is an abundance of sunshine for most months of the year. They can be grown in the Cape Peninsula but do not thrive there as they do in other parts of Southern Africa where the winters are dry and sunny. The species mentioned are hardy to frost and endure long periods of drought. They can be grown from seed or stem cuttings.

E. acanthocarpa TAMBOEKIE THORN

This is a handsome plant which grows wild near Queenstown. It stands fairly severe frosts and is grown in Johannesburg gardens. It has a large succulent root which helps it to endure long periods without water. The plant grows to 1 m (3 ft) and sometimes reaches twice this height. The branches have strong recurved thorns. The leaves are arranged in threes, each being as broad as it is long. They too have thorns, but small ones, at the back and on the leaf stalks. Even the seed-pods are armed with prickles. The colourful flowers are arranged in showy spikes. Each spike is about 18 cm in length with curved flowers. They are of a brilliant scarlet with tips of greenish-gold. The flowering time is spring. Some Bantu people use the seeds as a charm against evil whilst others use the pith of the root for making hats.

E. humeana DWARF KAFFIRBOOM

This species can be found growing in the eastern Cape, the Transkei, the warmer parts of Natal and in the eastern Transvaal. It reaches a height of 2-3 m (6-10 ft) and, like the others, is a deciduous plant; but, whereas the other species mentioned flower at the end of winter and in spring, this species flowers in summer. The leaves are composed of three broad leaflets with prickles on the back and on the leaf stalks. It bears large heads of handsome flowers of a rich and luminous scarlet colour. This species stands considerable frost.

E. zeyheri PRICKLY CARDINAL, PLOEGBREKER

Grows wild in the Orange Free State, Transvaal, Lesotho and Natal. The name of "ploegbreker" (plough-breaker) is derived from the fact that the huge root, which goes down about 1 m into the ground, is not easy to dislodge and resulted in the breaking of ploughshares when fields were being ploughed. It has large leaves with prickles and sends up stems to about 60 cm, on which large spikes of scarlet flowers appear in November. It is a handsome plant which stands drought and frost and it can be grown from seed fairly quickly.

Eroeda imbricata grows well in sandy soil and dry places.

Eroeda imbricata. Showing the flowers and foliage in greater detail.

Euryops *(Euryops abrotanifolius)* is a good shrub for spring colour.

EURYOPS DAISY BUSH, RESIN BUSH

DISTRIBUTION: There are many species of euryops in different parts of the country but most of the prettiest ones are native to the south-western Cape.

DESCRIPTION: These are evergreen plants with green or grey foliage and daisy-like flowers of golden-yellow which are held well above the leaves. They are apt to become leggy with age and new plants should therefore be planted out every four or five years. It is advisable also to trim the bushes back after they have flowered to keep them more shapely. They do well in coastal gardens and on the highveld.

CULTURE: They grow in any kind of soil but should be kept watered during the winter and spring. In gardens which experience severe frost it is advisable to plant them where the early winter sun will not strike the plants before mid-morning, or to plant them in the shelter of a wall or other plants. They grow quickly from seed and will flower in eighteen months from seed sown in late summer. Quicker results can be had by rooting cuttings.

This hybrid Euryops is one of the prettiest of the spring-flowering shrubs.

E. abrotanifolius
Grows to about 1.5 m (5 ft). The finely divided grey-green leaves are massed along the young reddish stems, and in September it becomes covered with bright yellow flowers which stand high above the foliage.

E. acraeus MOUNTAIN DAISY
This charming shrub has the specific name of *acraeus,* meaning "belonging to the summits" to describe its natural habitat in the Drakensberg, where it can be found growing at an altitude of approximately 2400 m. Although it has been grown in gardens in England for the last 20 years and has received an Award of Merit there, as far as I am aware no plants have been grown in private or public gardens in South Africa. It is a most decorative shrub suitable for gardens large or small. The leaves are small, only about 1.5 cm in length and 2 mm broad, but they are showy as they are of a luminous grey-green colour. In late spring and early summer the plant becomes covered with bright yellow daisies. It grows to a height of about 1 m (3 ft) and stands severe frost.

E. chrysanthemoides DAISY BUSH
(Gamolepis chrysanthemoides)
Can be found growing wild in the eastern Cape and Natal. It is a bushy plant reaching a height and spread of about 1 m (3 ft). It has attractively-cut leaves and produces masses of bright yellow flowers in winter and early spring. It is very quick-growing from seed or cuttings but is apt to become leggy with age unless it is trimmed back after flowering.

E. linifolia
Grows to 2 m (6 ft). The upper stems are clothed with soft, slender leaves only about 2 mm wide. The top part of the plant becomes covered with yellow daisies measuring 18 mm across. The flowering time is early spring.

E. pectinatus GREY-LEAFED EURYOPS
This neat plant grows to about 1 m (3 ft) in height and spread. The foliage is silvery-grey and arranged in rounded tufts. The leaves are attractively cut too. Although the flowers are smaller

 Tamboekie Thorn *(Erythrina acanthocarpa)* bears its splendid flowers in spring. (Page 107)

than in some other species it is one of the most decorative of the genus because of its pretty leaves. The flowering time is spring.

E. speciosissimus RESIN BUSH
(E. athanasiae)
Grows to 1.5 m (5 ft) and has finely divided slender leaves which give the plant a feathery appearance. The yellow daisies, 5 cm across, show up well above the foliage. It becomes leggy unless trimmed back after the flowering season. It flowers in late winter and early spring.

E. tenuissimus
A shrubby plant growing to 60 cm (2 ft) densely clothed with soft, thin, needle-like leaves. In spring it becomes smothered with little bright yellow daisies about 12 mm across. It is a splendid plant to have in a mixed flower border, or for edging a path or drive.

E. thunbergii
This species grows with cheerful abandon under dry conditions but needs water in winter and spring to encourage good flowering. The plant grows to 1 m (3ft) and in winter bears clusters of bright yellow daisies, each one about 1-2 cm across. It should be tried in dry gardens inland and in coastal ones too.

Euryops (unnamed hybrid)
One of the most attractive of the euryops is a hybrid which resulted from the crossing of *E. pectinatus* with *E. chrysanthemoides*. The plant which grows to 90 cm (3ft) has exceptionally attractive silver leaves, and in spring it becomes covered with bright yellow daisies which make a really splendid show against the silver colour of the leaves.

E. virgineus
Is a delightful species from the south-western Cape which makes a wonderful show of colour in mid-winter. The little flowers of bright yellow have a honey-like scent. It stands drought and frost.

GARDENIA AMOENA EAST LONDON GARDENIA
(G. gerrardiana)
DISTRIBUTION: Is to be found in the bush near East London.

DESCRIPTION: This is an attractive shrub which grows to a height of 2 m (6 ft) and bears dark green, glossy, oval leaves with slightly wavy margins, arranged opposite each other. The flowers which are decorative appear in late spring and early summer. Each flower consists of a slender tube which opens to a five-petalled, starry face. They are like wax, white in colour with suffusions of pink. Their rich perfume will scent the entire garden. They bear small, oval fruits.

CULTURE: This shrub does best in districts where winters are mild, although, once established, it will stand moderate frost. Plant it in rich soil in which there is plenty of compost, and water it regularly particularly during summer. In hot districts where the air is dry it should be planted where it is shaded for part of the day.

GREWIA OCCIDENTALIS ASSEGAI WOOD, KRUISBESSIE, CROSSBERRY, BUTTONWOOD
DISTRIBUTION: Its natural habitat is from the coastal parts of the eastern and western Cape and Natal, to the mountains of the eastern Transvaal, generally on the edge of forests.

DESCRIPTION: This is a decorative evergreen shrub growing to 2 m (6 ft) or more. It has fairly small, oval leaves with serrated margins and cyclamen-pink flowers about 3 cm across. The sepals of the flowers are of the same colour and almost the same length, so that the flower appears to have ten slender petals. The orange-yellow fruit which follows the flowers is eaten by some Bantu tribes. The flowering time is summer.

CULTURE: Make fairly large holes and fill them with plenty of compost and good soil. In hot inland gardens this plant should be set out where it will have some shade during the hottest part of the day. Water it well during the summer months. It tolerates mild frosts but needs protection in cold gardens. It can be grown from seed, but this takes a long time to produce flowering plants.

HERMANNIA HERMANNIA
DISTRIBUTION: There are many species of hermannia widely distributed through the country. Some decorative ones from the Cape Province are mentioned.

DESCRIPTION: These are quick-growing plants with flowers which are yellow, rose, cream, brick or purple. They are able to stand difficult growing conditions.

CULTURE: Although they survive long periods of drought and quite severe frost they do better if planted in good soil and given water regularly during their growing season.

H. althaeifolia — Dikbos

Occurs in the coastal bush near Port Elizabeth and in the south-western Cape. In nature it grows to about 60 cm (2 ft) usually in sandy ground. The leaves are widely oval and dark green, and the flowers are cup-shaped and bright yellow. It is a pretty shrub for coastal gardens. The flowering time is late spring.

H. stricta — Desert Rose, Rooi Opslag

Can be found in dry sandy soil in Bushmanland, where the rainfall is often less than 125 mm (5 in) a year. This is a striking little bush in September when it flowers. It grows to about 1 m (3 ft) and becomes enveloped by its rose-coloured flowers which have five rounded petals measuring 2-3 cm across.

H. vesicaria — Mustard Bush

Grows wild in the south-western Cape. It reaches a height of 45 cm ($1\frac{1}{2}$ ft) and has leaves which are oval and pointed. The bell-like flowers are only 12 mm long and pendulous. They are of a tawny-yellow colour. The flowering time is spring.

HOLMSKIOLDIA TETTENSIS — Mauve Chinese Hat Plant

(H. speciosa)

Distribution: Its natural habitat is the lowveld of the eastern Transvaal, and in warm areas further north.

Description: This is an erect-growing shrub to 2-3 m (6-10 ft) with long stems bearing clusters of flowers which are useful for arrangements. Each flower consists of a mauve calyx formed like a miniature Chinese hat about 18 mm across, and a corolla, like a little purple tube, hanging down from it. The flowering time is summer.

Culture: This large shrub grows quickly in any kind of soil but it does not tolerate cold. When cut back by frost, however, it frequently grows up again very quickly.

HYPERICUM REVOLUTUM — Curry Bush, Forest Primrose, Kerriebos

(H. leucoptycodes)

Distribution: Can be found in Natal, the north-eastern Transvaal and further north in Rhodesia, usually on the fringes of forests.

Description: It is an attractive evergreen shrub worth growing in gardens. It reaches a height of 2.5 m (8 ft) and bears arching branches with pleasing slender, pointed leaves which give off a rather pleasant smell of curry. The flowers are made up of five canary-yellow petals with an attractive tuft of stamens at the centre. They are carried in loose sprays towards the ends of the stems, and appear in the greatest number in spring, although it may have some flowers at other seasons also.

Culture: This robust shrub tolerates moderate frost and quite considerable drought when once it is established. Under temperate conditions it tends to become too large and it should be trimmed back each year after flowering.

IBOZA RIPARIA — Iboza

Distribution: Occurs in warm parts of Natal, Zululand, the northern and eastern Transvaal and Rhodesia.

Description: Iboza grows rapidly to a height of 2 m (6 ft) and has leaves which are soft in texture and heart-shaped, with serrated edges. The tiny mauve flowers are carried in long graceful sprays giving the whole plant a delightful misty appearance in late winter, when it flowers. It is a deciduous shrub and the flowers show up to perfection as they appear before the new leaves. The stems are useful for arrangements. The Zulus make an infusion of the aromatic leaves to relieve chest afflications.

Culture: Although it grows in any kind of soil the plant will grow faster and flower better if it is planted in soil to which some humus has been added. In hot inland gardens it should be planted where it has some shade during the hottest part of the day. Iboza cannot stand severe frost and in gardens where temperatures drop low in winter, it should be tried against a north-facing wall which traps the heat during the day and helps to keep the plant from being frosted at night. It grows readily from cuttings.

Iboza *(Iboza riparia)* bears misty mauve plumes of flowers in winter.

INDIGOFERA

DISTRIBUTION: There are more than two hundred different species widely distributed throughout the country.

DESCRIPTION: Indigofera includes plants which grow almost prostrate along the ground as well as small shrubs. They have small pea-shaped flowers of pink, rose, mauve or white. Some of them are useful for the flower border or to have growing in tubs on a patio.

CULTURE: They do not demand any special soil but many species prefer warm growing conditions to cold. They can be propagated from seed or cuttings.

I. cuneifolia

Grows at high altitudes in the eastern Cape and Natal. It reaches a height of 1.5 m (5 ft) and has leaves divided into three parts. The pea-shaped flowers are of a pleasant pinky-mauve colour arranged in spikes at the ends of stems. It is hardy to frost.

I. cylindrica

Grows to about 2 m (6 ft) or more and has little sprays of pale pink flowers in summer. The leaves are divided into many oval leaflets which are only 12 mm in length. It likes warm growing conditions and regular watering.

I. cytisoides

Is an erect-growing shrub to 2 m (6 ft). It bears small spikes of pink and rose flowers in autumn.

I. filifolia

Grows along sides of streams in the Cape Peninsula, and when grown elsewhere it should be watered well in winter. It reaches a height of 1.5 m (5 ft) and has soft, slender leaves and pink to rose flowers in autumn.

I. frutescens

A shrub to 2 m (6 ft) with pink flowers in late spring. Being native to the south-western Cape it requires water in winter.

JASMINUM JASMINE

DISTRIBUTION: Several attractive species are to be found in the warmer parts of the country from the coastal belt of the eastern Cape up into Natal and the warmer parts of the Transvaal.

Jasmine *(Jasminum multipartitum)*. A species with exceptionally attractive foliage and scented flowers.

DESCRIPTION: In their natural habitat these plants can be found scrambling over other shrubs, but when grown in gardens they can be trained to serve the purpose of a true climber. When trained along a fence jasmine makes a fine dense hedge, or it can be trained up a wall. It can also be treated as a shrub and clipped to keep it within a limited space. They look effective, too, when grown in large pots or tubs on a patio. They have decorative, dark green, rather shiny leaves, broader at the base than the apex. The buds are suffused with pink and the open flowers have waxy, white petals and a sweet perfume.

CULTURE: Although the most decorative species occur naturally in regions where winters are mild and the rainfall good, they will stand fairly severe frost and a good deal of drought, when once they are established. In hot inland gardens they appear to do best when the ground around their roots is shaded by other shrubs or trees.

J. angulare EAST LONDON JASMINE

This twining shrub grows in the bush near East London and through the Transkei into Natal. It has angled stems and trifoliate leaves. It bears

Lachnaea *(Lachnaea buxifolia)* is a charming shrub for gardens large and small.

sweetly-scented flowers for many months of the year, but particularly in spring and early summer. It will scramble or climb up through trees to a height of 6 m (20 ft). In the open it may remain shrub size. When this plant flowers well the Xhosa people take it as an indication of a good season ahead.

J. breviflorum NATAL JASMINE
(J. gerrardii)
Occurs between Durban and Pietermaritzburg, and elsewhere in Natal and near East London. The flowers are about 18 mm across and have a rich fragrance. It bears flowers in late winter and early spring and at other times of the year, too.

J. multipartitum JASMINE
This is the indigenous species most often seen in gardens. It has larger leaves and flowers and is altogether more handsome than the other species mentioned. It makes a fine background to other plants and the sweet scent of its charming flowers adds another dimension to the garden scene. It flowers from late winter to early summer. The scandent stems may grow to many feet in length and it is a good plant to grow along a fence or train along a wall, or it may be clipped back to keep it as a shrub.

LACHNAEA DENSIFLORA LACHNAEA, BERGASTER

DISTRIBUTION: Grows naturally in the south-western Cape.

DESCRIPTION: This is a shrub to 45 cm ($1\frac{1}{2}$ ft) with great charm, suitable for large or small gardens. It has soft slender stems tightly clothed with little needle-like leaves about 12 mm long which point upwards and hug the stem. The tiny flowers have four starry petals and are very attractive. They are tightly clustered together in rounded heads like miniature posies. Each flower is ivory to white in colour with a flush of pink occasionally at the centre. The flowering time is September. *L. buxifolia* is another species from the western Cape worth trying, and *L. diosmoides* is a pleasing species found near Port Elizabeth.

CULTURE: Lachnaea do well in poor soil but grow and flower better when planted in soil to which compost or manure has been added. The plants should be watered in winter to encourage good flowering. They stand moderate frost and fairly long periods of drought.

LEBECKIA LEBECKIA, WILD BROOM, WILDEBESEMBOS, FLUITJIES

DISTRIBUTION: The two species worth cultivating in the garden come from the south-western and western Cape, where they are sometimes found growing wild in regions where the summers are extremely hot and dry and where the total annual rainfall may be less than 500 mm (20 in).

DESCRIPTION: These shrubs are of exceptional beauty in spring when they bear masses of yellow pea-shaped flowers in graceful spikes. They look effective in a flower border or interplanted with other shrubs.

CULTURE: They will stand long periods without water from spring to autumn and are good plants for hot, dry regions. They need to be watered in winter to encourage good flowering. They grow in any kind of soil and stand quite considerable frost.

L. cytisoides WILD BROOM, WILDE BESEMBOS
This species is very like Spanish broom in appearance but the habit of growth of the plant is more graceful and the foliage is prettier. It grows to about 1.25 m (4 ft) in height and more across,

and has slender leaves arranged in threes. In early spring it bears long sprays of sparkling, yellow, pea-shaped flowers. It grows from the dry areas of Namaqualand south to Malmesbury.

L. simsiana DWARF LEBECKIA
This is a low-growing species which sprawls along the ground spreading across 1.5 m (5 ft). It has long, soft, needle-like leaves which are closely arranged along the arching stems, and in late winter to mid-spring it bears handsome spikes of bright yellow flowers. It can be found growing in sand, gravel or clay soil, in areas with a very low rainfall and also where there is a high winter rainfall. This species is a delightful plant for the rock-garden.

LEUCADENDRON

DISTRIBUTION: There are thought to be about ninety species, nearly all of them from the winter-rainfall area of the south-western Cape.

DESCRIPTION: These handsome plants which are usually medium to large shrubs, form an important group of the protea family. Many of them are very decorative in the garden and the long-stemmed flowerheads make large and splendid

Wild Broom *(Lebeckia cytisoides)*. Both leaves and flowers are decorative.

Dwarf Lebeckia *(Lebeckia simsiana)*. A delightful shrublet for a rock-garden.

floral decorations. They last well in the vase and even after their fresh beauty has faded many species provide interesting material for dried arrangements.

The male and female flowers are carried on separate plants, interesting in their diversity and often confusing to the non-botanist who may think they are different species. The male flowers have a fluffy appearance; the females form a cone surrounded by a circlet of leaves, which in some species change from normal shades of green to rich yellows and reds.

CULTURE: Like other members of the protea family most leucadendrons thrive in soil which is distinctly acid. In areas where the soil is not acid it is advisable to make holes 60 cm (2 ft) wide and deep and to fill these with acid compost and leaf mould, or some peat, which helps to retain the water supplied to them as well as providing an acid base. They can stand fairly severe cold but only if they are watered well and regularly during the autumn to spring period. In their natural habitat many species are dry throughout the summer months, and, when they are grown where heavy summer rains are usual, it is advisable to ensure that they have good drainage. They can be grown from seeds or cuttings, but gardeners are advised to purchase plants from a nursery, as growing them from seed takes a long time to produce flowering plants, and rooting them from cuttings is not easy.

L. album
(L. aurantiacum)
This species reaches a height of 1.25 m (4 ft) and has silvery leaves and cones. The foliage is needle-like. The male plant bears fluffy yellow flowers in plumes. It does well in coastal gardens and also stands quite severe frost. It grows in alkaline as well as in acid soil.

L. chamelaea
(L. decurrens)
This is a pretty species which grows to about 1 m (3 ft). The male flower is very attractive in spring with its fluffy dome of lemon-yellow flowers surrounded by leaves of a paler hue—like a halo.

L. comosum
(L. aemulum)
In this leucadendron the female flower is decidedly showier than the male, although both are worth having. The plant occurs in nature along mountain slopes from Paarl to near Port Elizabeth. It grows to 1.25 m (4 ft) and is at its best in spring and early summer. The fluffy brown male flowers are surrounded by slender sulphur-yellow leaves. The female plant bears a long cone of tan to deep brown which stands out in sharp contrast to the curving yellow leaves which form a bowl around it. The leaves lower down on the plant are very slender almost like those of a pine.

L. conicum
Although the plant grows to 3 m (15 ft) the flowerheads are not large. They measure only 3-4 cm across but are carried in large numbers and are decorative. The fluffy yellow centre of the male flower shows up well against the surrounding crimson bracts.

L. coniferum
(L. sabulosum)
This is widely distributed in the Cape Peninsula where its stems are used in quantities by florists for winter arrangements. The plant grows to 1.5 m (5 ft) and more. The female plant has oval cones shaded with pink. It appears to do well in alkaline or acid soil.

L. daphnoides
Grows wild on slopes near Paarl and Franschhoek. This is a decorative species with large heads and yellow involucral leaves beautifully suffused with crimson. It grows to about 1 m (3 ft) in height.

L. discolor
This is one of the most robust and one of the prettiest of the leucadendrons, specially when seen in masses against a hillside. It reaches a height of 2 m (6 ft) and has erect stems which end in broadly oval decorative leaves arranged in cup-formation like a half-opened water lily. They are yellow often tinged with red in winter. The male flower is prettily coloured yellow and red and the female cone is very showy, being green suffused with red and yellow. This is a fine species for arrangements as well as for garden show.

L. elimense
Is to be found in the Bredasdorp district and is suitable for gardens small and large. It grows to about 1 m (3 ft) and is at its prettiest in spring. The male flowerhead is bright yellow surrounded by a halo of pale yellow leaves. In the female, the leaves surrounding the cone and the cone itself often turn a rich shade of red.

Leucadendron daphnoides (female)

Leucadendron discolor (male)

Leucadendrons make a handsome show in the garden and provide long-lasting material for arrangements.

Leucadendron discolor (female)

Leucadendron eucalyptifolium (male). Is graceful in form and foliage.

Leucadendron floridum. A charming species for the small garden.

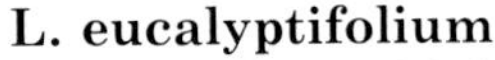

L. eucalyptifolium
A tall species which grows to 2 m (6 ft) and has graceful stems of leaves which become coloured sulphur-yellow at the top in winter and spring.

L. floridum
This is one of the most charming of the leucadendrons not only for the garden but also for arrangements. It grows to about 1.25 m (4 ft) in height and spread and has slender silvery leaves 2-3 cm long and 6 mm wide. It flowers profusely producing decorative flowerheads clustered together. Each little flowerhead is delicately tinted pale lime-green and yellow, and surrounded by a halo of velvety, biscuit-yellow and silvery-green leaves. It is at its prettiest in early spring.

L. galpinii
This is a species for the large garden or for the collector. It grows to 2 m and is upright in habit. The male flowers form creamy-yellow pompons and the female plant bears rounded grey cones with a flush of pink.

L. gandogeri
(*L. guthrieae*)
Reaches a height of 1.5 m (5 ft) and bears creamy-yellow male flowers surrounded by bright yellow leaves sometimes rimmed with pink at the edges. The female flowerhead has a long cone surrounded by leaves which are yellow edged with pink and rose.

L. laureolum
(*L. decorum*)
This species which is found from Cape Town to Caledon is not as showy as some of the other species and the female plants grow too large for the garden of average size. They reach a height of 2.5 m (8 ft) and have yellow cones surrounded by yellow leaves. This is a useful plant for roadside or park planting.

L. loranthifolium
Grows to 1.5 m (5 ft) and is spreading in habit. The new stems are tinged with red and the oval pointed leaves are about 4 cm long and prettily coloured a metallic blue-green shade. The female flowerhead looks like a large, ornamental, rounded button of glistening green and dusky maroon.

L. macowanii
This rare species is suitable only for the large garden or park. It is not particularly decorative but makes a good background plant. It grows to 3-4 m (10-13 ft) and has slender leaves of dark green. It differs from other leucadendrons in that the leaves surrounding the flowerheads do not

change colour, but remain green throughout the year. The male flowers are bright yellow and the cones carried by female plants turn russet-red when mature.

L. microcephalum
(*L. stokoei*)
Grows to 1.25 m (4 ft) and is compact in form. In winter when the leaves at the top turn yellow it makes the fields where it grows glow with colour. In this species the flowerheads are surrounded by a ring of brown bracts within the yellow terminal leaves.

L. modestum
This is a good species for the small garden as it seldom grows taller than 30 cm (1 ft) in height and spread. In spring the terminal leaves turn a charming shade of lemon-yellow which makes a good contrast to the golden-yellow colour of the male flowers. The flowerheads are only 3-4 cm across and carried in profusion. They are fine for arrangements as well as making a show in the garden.

Leucadendron discolor and *L. loranthifolium* (metallic-blue) provide flowers for long-lasting arrangements.

The female flowerhead of *Leucadendron platyspermum* is unusual in form and colour.

L. muirii
Grows to 1 m (3 ft) and its attraction is in the large silvery cones of the female plants. It is to be found near Bredasdorp and may grow better in alkaline soil than do other species.

L. platyspermum KNOBKERRIE BUSH
This species grows to 2 m (6 ft) and is upright in habit. The male plant does not produce showy flowers, but the female plant is pretty in the garden and bears cones which are woody and grooved and which make interesting everlasting flower arrangements. The terminal leaves surrounding the female cone are misty yellow for a time in spring. All the leaves of the male plant turn yellow. In this species the leaves near the top of the plant are more slender than those lower down.

L. nervosum
Is a tall plant reaching a height of 2 m (6 ft). The flowerheads are not brightly coloured but interesting. The terminal leaves are shaded in soft tones of yellow, green and brown and covered with fine cobwebs of woolly hair.

L. pubescens
(*L. seriocephalum*)
Grows to 1.5 m (5 ft) and has oval leaves 2-3 cm long facing upwards and closely arranged along the top stems. The pale yellow flowers on the male plant are surrounded by leaves which turn yellow. The female plants bear silver cones.

L. rubrum
(*L. plumosum*)
This species is common in the south-western Cape spreading east towards Humansdorp. It has grey-green leaves and, in spring the male plants bear masses of tiny flowers of sulphur-yellow in clusters along the ends of the stems. The female plant has oval cones surrounded by grey-green leaves which provide fine material for dry arrangements. The plants grow to 2.5 m (8 ft).

L. salicifolium
(*L. strictum*)
A tall species to 2 m (6 ft) with yellow involucral leaves. The female plant bears an oval cone which is suffused with purple.

L. salignum
(*L. adscendens*) GEELBOS
This is a pretty shrub 60 cm (2 ft) in height, to be found growing near the coast from Cape Town to Port Elizabeth. During winter and spring the leaves around the flowers are of pale lemon-yellow sometimes edged with crimson, and in some areas they are prettily tinged with apricot and red at this time of the year. It provides colour in the garden and fine stems for arrangements. It is hardy to frost.

Leucadendron sessile (male). Becomes attractively coloured in late winter and early spring.

L. sessile
(*L. venosum*)
This handsome species grows to 1 m (3 ft) and has involucral leaves which may be yellow or crimson, depending on where it is growing. It is a decorative plant for large or small gardens.

L. spissifolium
(*L. glabrum*)
Grows to 1.25 m (4 ft). Both the male and the female plants bear attractive flowerheads, but those of the female are the prettier of the two. The male flowers are green when young, turning yellow as they mature, and are surrounded by broad, pale-yellow leaves forming a cup or bowl. In the female flower the russet cone is surrounded by overlapping bracts of creamy-yellow often tinted with crimson.

L. tinctum
(*L. grandiflorum*)
This is a very handsome plant in winter and early spring. Both the male and female flowers are attractive. The male flowerhead is yellow and fluffy and surrounded by colourful leaves of bright rose and yellow. The female flowerhead is a maroon-red cone with a halo of leaves prettily coloured deep rose at the base and lime-green towards the tips. The plants grow to about 1.25 m (4 ft).

L. uliginosum
Is a useful and decorative plant because of the silvery tint of its leaves. The flowers are not showy but the stems of leaves are useful for arrangements as well as being decorative in the garden. It grows to 2 m (6 ft) and is resistant to fairly severe frost.

L. xanthoconus
(*L. salignum*) GOLD TIPS
This is a pretty shrub in late winter and early spring when its leaves, particularly those towards the top of the plant, turn bright yellow. It grows to about 2 m (6 ft) and is a splendid plant for large gardens and for providing material for arrangements.

LEUCOSPERMUM PINCUSHION, LUISIES
DISTRIBUTION: The greatest concentration of leucospermums is on mountain slopes in the south-western Cape.

DESCRIPTION: Some of the leucospermums are the

Leucadendron tinctum (female). This is a decorative species for cool gardens.

Leucadendron tinctum (female). Close-up of flowerhead.

Leucadendron tinctum (male) has flowerheads of subtle shades. ▷

most rewarding of all our native shrubs. They have good foliage, a pleasing habit of growth and outstanding flowers which last well on the plant and when picked for arrangements. The different species vary in size considerably, from small prostrate ones to tall ones of 3 m (10 ft). They belong to the protea family but the flowers are different in form from those of the protea. The leaves are different too. They usually have three to five notches at the apex, whilst those of the protea do not have such indentations. The flowers are clustered together in a compact rounded head, and the styles which protrude in the open flower have enlarged ends like a pinhead, giving the flowerhead the appearance of a pincushion full of pins, which accounts for the common name of "pincushion". The other common name of "luisies" refers to the seeds of some species which look like ticks.

CULTURE: In their natural habitat most of these plants grow on slopes where the soil is acid and the rainfall is high during autumn to spring. Even in the long, dry Cape summers they have a certain amount of humidity, not only because of their proximity to the sea but also because of the moisture often present in mist on the slopes where they grow. When grown inland they should be watered regularly and well from autumn to spring. In regions where the summer rainfall is very heavy it is advisable to set the plants out where the drainage is good—i.e. on sloping ground. If there is no slope to provide drainage make extra large holes and fill the bottom of the holes with rubble. When planting leucospermums where the soil is poor or alkaline make holes 60 cm deep and across and fill them with acid compost mixed with some of the soil previously removed, as they do not thrive in alkaline soil. Leucospermums will stand a good deal of frost but very severe frost may damage the plants particularly when they are young. In regions where sharp frosts occur it is advisable to plant them where they have some protection, and where they are shaded from the early morning sun until 9-10 a.m. Young plants can be protected from frost damage at night by having a cardboard carton inverted over them each evening. This must be removed during the daytime as members of the protea family like to have a free circulation of air about them. For this reason it is inadvisable to set out plants near the walls of buildings or to cover them with permanent winter protection in the form of straw or plastic "tents".

In the summer-rainfall region it is advisable to plant them out at the beginning of the rainy season, i.e. from September to November, whilst in the winter-rainfall region it is best to plant them from March to August. They can be grown from seeds, but this takes three to five years to produce flowering plants and most gardeners therefore prefer to purchase plants from a nursery. They transplant better when small and, if the plants received from the nursery in tins or plastic containers appear smaller than other shrubs in containers, do not think that the nurseryman is sending you immature plants. He is aware that young plants transplant more readily than large ones. The height and spread given below are under average conditions. Under optimum conditions they may grow larger than indicated.

L. bolusii
(L. album) WHITE PINCUSHION
Grows to 2 m (6 ft). This is not as decorative a plant in the garden as the species with larger flowers, but it is useful for flower arrangements. The flowers, which give off a honey-like scent, are carried in heads about 18 mm across. These heads, or little pincushions, are arranged several together in clusters at the tops of the stems. The flowerheads are ivory tipped with grey. The leaves are oval and 3-5 cm long. The flowering time is late winter and early spring.

L. calligerum
(L. puberum)
This is not a showy plant but it produces charming little heads of flowers for arrangements. The bushes are robust in growth and the flowerheads are 2-3 cm across with an overall colouring of pink, rose and silvery-grey in the bud, with ivory showing up also on open flowers. It flowers in spring to early summer.

L. catherinae CATHERINE WHEEL
Grows to 1.5 m (5 ft) with a spread of as much or more, and has large and very attractive flowers which open up to look rather like a Catherine wheel about 10 cm across. It is a very pleasing plant both when in bud and when the flowers open. The flowers are a soft apricot yellow shade flushed with rose at the tips. The main flowering time is from September to October. The leaves are 5-7 cm long, oval and greyish-green in colour.

Pincushion *(Leucospermum cordifolium)*. A well-grown specimen may carry three hundred or more flowers at a time.

Leucospermum calligerum has softly shaded flowers which look charming in arrangements.

Pincushion *(Leucospermum cordifolium)*. Showing detail of flower.

The flowers of the Catherine Wheel *(Leucospermum catherinae)* are most decorative.

L. conocarpodendron KREUPELHOUT
(L. conocarpum)
This is a large species which grows to 3 m (10 ft) in height and spread, but it can be kept trimmed back to smaller size by cutting off the faded flowers with a long piece of stem. The common name, in use since 1680, is a Netherlands word meaning "shrubby undergrowth used as firewood". In the early days of the settlement at the Cape the bark of this plant was used in large quantities for tanning. It is not surprising, therefore, that this robust plant which once grew in masses all over the Cape Peninsula and further afield, is now rarely seen. It is interesting to note that it was grown in Holland early in the eighteenth century. Its flowers are outstanding although they are sometimes partially hidden by the large leaves. The flowerheads measure 10 cm across and form a rounded dome of gleaming, golden-yellow. The leathery leaves are 10 cm long and 4 cm across. This plant is suitable for the large garden and park and for edging a long drive. It is at its best from September to November.

L. cordifolium PINCUSHION,
(*L. nutans*) SPELDEKUSSING
This is one of the most decorative species and the most rewarding both in making a fine show in the garden and in producing flowers for arrangements. It reaches a height of 1-1.25 m (4-5 ft) and spreads across much more than this, as the stems tend to curve out and up making the whole plant rather elegant in form. A well-grown plant may spread across 4 m (13 ft). The flowerhead is attractive at all stages. It forms a high dome when immature, and as it matures the glistening styles with their "pinheads" project gracefully in a curving fashion. The colour ranges from apricot to coral and rose. Their main flowering season is late winter and early spring. A mature plant may carry as many as three hundred or more flowers at one time. The flowers last for a long time on the plant and in arrangements. The plants tend to become woody and may die off in eight to nine years, and it is therefore advisable to plant new ones every few years.

L. cuneiforme
(L. attenuatum)
Grows to 1.5 m (5 ft) and is tolerant of a variety of soils. It can be found growing naturally as far east as Grahamstown. The stems are pale grey and hairy and the leaves are leathery, square across the top and clearly notched at the ends. They overlap each other neatly. The flowerheads are about 5 cm across and show up well against the foliage. They are first yellow, apricot or orange in colour and later turn to flame. The flowers invariably grow in pairs at the ends of stems. The flowering time is winter and early spring.

L. glabrum
Is a robust plant growing to a height of 2 m (6 ft) with exceptionally showy flowerheads in August and September. Each one is a high-domed arrangement. In bud, silky grey hairs clothe the middle of the inflorescence and, when the flowers mature, the gleaming styles stand firmly erect. The central part then becomes crimson which shows up well against the silky hairs. The styles of amber to salmon are uniquely tipped with spear-shaped ends of pale lime-green, rose and yellow. It has leathery leaves 7-10 cm long and 2-3 cm across at their widest, which is the apex. They are deeply and clearly notched at the apices, and the tops of the leaves are tinged with maroon-red.

L. grandiflorum RAINBOW PINCUSHION
This is a quick-growing species which makes a handsome show at the back of a border. It bears flowerheads which measure 5 cm across, with styles which grow rather erect. The flowers change from yellow to orange and red as they mature making the whole plant look gay with many colours at the same time. It grows to 2 m (6 ft) and is erect in habit. The main flowering time is early spring.

L. gueinzii
This is a rare species found in the Jonkershoek mountains. It bears a resemblance to *L. catherinae* but the flowerheads are not as attractive either before or after opening. The foliage is greyish-green and the styles of the flowers turn over outwards at the tips. They are sulphur-yellow at the base and crimson along the upper part. They appear from September to November.

L. lineare NARROW-LEAFED PINCUSHION
The flowers of this species are very like those of *L. cordifolium* but its leaves are different. They are long and slender, rather like those of a pine, and carried close along the stem, giving the stems a rather feathery appearance. The flowerheads vary in colour: pale pink, coral, green. The main flowering season is early spring. The plant grows to 1.25 m (4 ft) in height and spreads across double this.

◁ Kreupelhout *(Leucospermum conocarpodendron)* is a robust plant with magnificent flowers.

L. muirii MUIR'S PINCUSHION

This species grows to 1.25 m (4 ft) and is a charming plant for the small garden as well as the large one. The greyish-green leaves are oval and narrow with little waxy red blobs marking the notched apex of each leaf. The flowers stand out well above the foliage and are up to 4 cm across. The yellow styles curve out attractively and the corollas roll back and look like slender rolls of red ribbon with delicate yellow stripes. The flowering time is from July to October.

L. mundii

This species can be found in the Langeberg and on Garcia's Pass, Riversdale. It grows to 1 m in height and spread and has leaves which are deeply notched at the apex. It is a plant of fine form and particularly decorative during early spring when it is in full flower. The flowers change in colour from yellow through orange and finally become red, and a plant covered with flowers at different stages of maturity makes a really striking show. It is similar to *L. oleifolium*.

L. oleifolium TUFTED PINCUSHION

(*L. crinitum*)

Grows to 1 m (3 ft) or more and has hairy grey-green leaves 6 cm long and 18 mm across. The upper leaves are deeply notched. The flowerheads are generally carried in twos or threes at the ends of stems. They are tufty in form and change in colour as they mature from yellow to orange-red. This species makes a colourful show in spring when it is covered with open and half-open flowers. It stands considerable frost.

L. prostratum CREEPING PINCUSHION

This is a delightful plant to have hanging down over a bank or rock wall. It grows flat along the ground and has masses of slender little green leaves which make a carpet. In spring it becomes covered with its little "pincushions" measuring 4 cm across, coloured from yellow in the newly opened flowers to apricot and rose as they mature.

L. reflexum ROCKET PINCUSHION

This is a spectacular plant for it has exceptionally handsome flowers and attractive foliage too. The plant grows to 3 m (10 ft) in height and almost as much in spread and has gracefully curving stems. It is a striking plant even when not in flower, for the small silvery-grey leaves are neatly arranged along the stems. In late winter and early spring it becomes a magnificent sight, crowned with its large showy heads of flowers which are of coral to tomato-red, with tints of silver from the fine hairs which clothe the flower. A well-grown plant is said to be able to produce a thousand blooms in a single flowering season. The flowers do not last long, but as there are always new ones opening, the bush is a fine sight for a long time. The common name is derived from the way the bright red styles reflex and project downwards giving the flower the appearance of a rocket taking off.

L. rodolentum

(*L. candicans*)

This species is for the large garden or park only. It is not as attractive as some of the species with large flowers. It grows to 3 m (10 ft) and spreads across much more. The grey-green oval leaves are 4 cm in length and hug the stem giving the plant a pleasing appearance. The flowerheads are only 2-3 cm across, golden-yellow and carried in clusters of about four. It flowers in spring.

L. tottum FIRE WHEEL PINCUSHION

Grows to 1 m (3 ft) and has narrow, oval, pointed leaves about 5 cm long. The flowerheads measure 7 cm across and have a domed centre. The colour of the flowers is a charming and subtle shade of dusky pink with crimson to maroon tips to the styles. It flowers later than other species, being at its best from October to December. This is a very lovely species for the garden or for growing in tubs.

L. truncatulum

(*L. buxifolium*)

This species grows in limestone country near Bredasdorp and is therefore recommended for gardens where the soil is alkaline. The flowerheads which appear in spring are very similar in appearance to those of *L. muirii*, but the flowers differ in having hairy perianths and longer styles, and the leaves are a little larger than those of *L. muirii*.

L. vestitum UPRIGHT PINCUSHION

(*L. incisum*)

This is a good-looking, erect-growing plant to 1.25 m (4 ft). The leaves are neatly overlapping, about 6 cm long and 2-3 cm wide. They are more deeply notched at the apex than most of the other species described. The gleaming flowerheads measure 12 cm across and are coral-red in colour with styles of rose or yellow. It flowers from mid-spring to early summer.

◁ *Leucospermum glabrum* holds its beautiful blooms proudly erect.

The flowers of the Narrow-leafed Pincushion *(Leucospermum lineare)* are long-lasting on the plant and in arrangements.

The Tufted Pincushion *(Leucospermum oleifolium)* is a decorative plant for the garden or a container.

Creeping Pincushion *(Leucospermum prostratum)* makes a fine show hanging over a wall or bank.

Rocket Pincushion *(Leucospermum reflexum)*. A lovely species for a large garden.

Leucospermum rodolentum. A species for the large garden or park.

Muir's Pincushion *(Leucospermum muirii)*. This is a delightful leucospermum for gardens large or small.

Leucospermum mundii is a robust species with unusual heads of flowers.

Rocket Pincushion *(Leucospermum reflexum)* showing manner of growth. The coral flowers show up beautifully against the grey of the leaves.

Leucospermum glabrum and *L. truncatulum* on left. Species on right not named.

Mountain Dahlia *(Liparia splendens)* has handsome flowers in spring.

Firewheel Pincushion *(Leucospermum tottum)* bears its pretty flowers later in the year than the other species.

LIPARIA SPLENDENS MOUNTAIN DAHLIA,
(L. sphaerica) GEELKOPPIE

DISTRIBUTION: Grows on the slopes of mountains in the south-western Cape as far east as Riversdale.

DESCRIPTION: When in flower this plant is eye-catching but to grow it well is not easy. It reaches a height of 1.25 m (4 ft) and is evergreen. In spring it bears large, rounded, nodding heads of flowers which measure about 8 cm across. Each individual flower is pea-shaped and golden-yellow flushed with terra-cotta, the whole head being surrounded by large, oval brown bracts.

CULTURE: Plant it in soil in which there is plenty of acid compost, and water it well throughout the winter. If grown in a part of the summer-rainfall region where the rainfall is heavy it is advisable to see that it has good drainage. The plant can be grown from seed or cuttings, but it is not easy to raise plants, and gardeners are advised to purchase plants rather than endeavour to raise them.

LOBOSTEMON FRUTICOSUS EIGHT-DAY-HEALING BUSH

DISTRIBUTION: Its natural home is on the slopes and plains of the south-western Cape.

DESCRIPTION: The common name remains a mystery as I have not been able to discover what malady it is supposed to heal within eight days. This is an evergreen shrub growing to a height of 1 m (3 ft) with a similar spread. It is particularly pretty when young. Old plants should be discarded when they become leggy and new ones planted. The young stems are rose-coloured and show up through the leaves which are oval, 2-7 cm long, 6-12 mm broad, and pitted with small spots. The flowers which appear in August and September are very lovely. The buds are palest shell-pink and open into funnel-shaped flowers of a pastel powder-blue colour flushed with rose in the throat. In some areas the flowers are of a most alluring shade of blush-pink or pale rose. They are carried in little clusters. *L. trigonus* which grows on sandy plains near Port Elizabeth is somewhat similar. It grows to 30 cm (1 ft) and flowers a little later, from October to December. *L. trichotomus* is another attractive species from near Piketberg. It grows to about 30 cm and has pretty pink buds which open to charming clusters of powder-blue flowers in October.

CULTURE: Although it grows well in any kind of soil, like many wild plants, it performs better if planted in soil which has been improved by the addition of some compost. It can be grown from seed sown in spring or summer, or from cuttings. It should be watered during autumn to spring and may be left dry throughout the summer. It is tender to severe frost and is best suited to gardens where frosts are never more than mild.

MACKAYA BELLA MACKAYA

DISTRIBUTION: Occurs in Natal and in warm forest regions of the north-eastern Transvaal.

DESCRIPTION: The plant grows to 2 m (6 ft) but can be kept cut back to much smaller size. The flowers are funnel-shaped, opening to a face made up of five broad and pointed petals, the lower three projecting out and the upper ones being more upright. They are of a delightful pale mauve colour with delicate veins of darker hue. The plant is an evergreen, and attractive throughout the year because of its glossy leaves. It can be used to brighten the ground under tall trees. It flowers in October and November.

CULTURE: This shrub does best in shade, particularly when grown in inland gardens, but it can be grown in the open in coastal gardens. Plant it in soil to which plenty of compost has been added. It is tender to frost and in cold gardens should be planted where it has protection under trees, framed by other plants or near a wall. It grows quite easily from cuttings.

MELIANTHUS HONEY FLOWER, KRUIDJIE-ROER-MY-NIE

DISTRIBUTION: Species of this plant can be found in districts north of Cape Town and east to the Transkei, Natal and up north into the Transvaal. The name is derived from the Greek *meli* (honey) and *anthos* (flower) as the flowers secrete an abundance of nectar.

DESCRIPTION: Melianthus are worth growing for their decorative foliage, although the flowers of some species are distinctive too. They are quick-growing and tolerant of poor soil and some neglect. They can be used as accent plants here and there in a garden or to line a drive.

CULTURE: They make speedy growth when planted in good soil and watered regularly; once established they stand fairly long periods without water. They can be grown from seed or root sections.

M. dregeana KRUIDJIE-ROER-MY-NIE

A little shrub common in parts of the Transkei. It grows to 1 m (3 ft) or more, and has large com-

Lobostemon trichotomus seems to thrive in poor soil.

Eight-day-healing Bush *(Lobostemon fruticosus)* is a delightful sight in late winter.

Mackaya *(Mackaya bella).* A pretty shrub for a shady place.

pound leaves made up of oval, pointed leaflets which clothe the plant luxuriously. In late spring it bears coral-red flowers which are generally partially covered by the leaves. The common name is derived from the fact that the plant gives off an unpleasant smell when handled.

M. insignis
Is a species found in the Transvaal which will stand fairly severe cold. It is at its best in spring and summer.

M. major LARGE HONEY-FLOWER
This plant with decorative leaves is found from Clanwilliam south to Cape Town and east towards Swellendam. It has very large grey-green leaves which are beautifully divided into leaflets with deeply serrated and ruffled edges. It grows to 1.5 m (5 ft) with thick stems coming up from the base. In spring it bears spikes of flowers which are 30 cm (1 ft) or more in length. The flowers are covered by brown bracts which show up well against the colour of the foliage, and later it bears large pale green seed-pods. If it is cut back by frost it grows rapidly again, but in frosty gardens it is advisable to plant it in a sheltered position. The plant gives off an unpleasant smell when cut. It makes a fine accent plant.

M. minor DWARF HONEY-FLOWER
Grows to 1 m (3 ft) and has attractive leaves of grey-green. The flowers are partially hidden by leaves and do not make much of a show but the plant is worth growing for its decorative leaves.

METALASIA METALASIA
DISTRIBUTION: Different species grow along the roadsides and mountain slopes from the western to the eastern Cape and in the Transvaal.

DESCRIPTION: The species best for garden show are shrubby in habit of growth to 1 m (3 ft) or more, with woody branches and heath-like leaves carried in tufts along the stem. The flowers are very small and densely packed together in round heads. The colour varies with the species, and some of them, apart from being decorative in the garden, yield flowers which last for a very long time in arrangements. Metalasia does well in coastal gardens.

CULTURE: Metalasias grow in any kind of soil—sand, gravel or clay— and once established they stand long periods without water, but they undoubtedly do better if given some attention. They stand moderate frost.

The Large Honey-Flower *(Melianthus major)* has decorative leaves and handsome spikes of flowers in winter.

M. aurea
This little shrub grows well in stony ground near Port Elizabeth. In winter it becomes covered with heads of tiny flowers. Each head is 2-3 cm across and their bright yellow colour makes the whole bush a cheerful sight.

M. muricata BLOMBOS
Grows to 1.5 m (5 ft) or more, and flowers in early spring, if watered during winter, and in summer if it is left dry during the cold months of the year. It is hardy to cold and drought and bears attractive heads of little flowers—each head measuring 4-5 cm across. The flowers which are white, ivory and grey look most effective in arrangements and they last for weeks.

M. rhoderoides
A neat and attractive shrub of merit for the garden because of its silver-grey colour. It grows to 45 cm ($1\frac{1}{2}$ ft) in height and spread, and has spiky little silver leaves rather like those of an erica in shape, carried in tufts close together all along the grey stems. It looks most attractive near blue flowers. The stems last well in water.

Mauve Metalasia *(Metalasia seriphiifolia)* bears attractive flowers which last for weeks in arrangements.

Silver-leaf Mimetes *(Mimetes argenteus)* is a rare and beautiful plant.

Mimetes cucullatus is most colourful in late winter and early spring.

Red and Yellow Bottle Brush *(Mimetes hirtus)*. This species should be kept moist throughout winter.

M. seriphiifolia MAUVE METALASIA

This is a showy bush which can be seen in dry areas north of Cape Town and also east of the Cape Peninsula. It grows to 1 m (3 ft) or more and bears quantities of flowerheads measuring about 2-4 cm across and coloured a pleasing shade of pale magenta. They are decorative for weeks on the plant and last for a long time when picked. It should be watered during winter and early spring. The flowering time is from August to December. It is hardy to frost.

MIMETES MIMETES, SOLDAAT

DISTRIBUTION: Occur in the south-western Cape particularly in the region between Betty's Bay and Caledon.

DESCRIPTION: These are charming evergreen shrubs which make a wonderful show on the hill-slopes where they grow wild. Their decorative value is not due to the flowers alone, which are hardly seen, but also to the colour of the leaves on the upper stems.

CULTURE: Mimetes require the same kind of growing conditions as proteas—i.e. well-drained, acid soil. To ensure good growth make holes 60 cm (2 ft) deep and across and put in plenty of acid compost. In cold gardens they should be protected from frost when young. They must be watered well and regularly from autumn to spring when they flower. During the summer they can do with little water but should not be allowed to become completely dry.

M. argenteus SILVER-LEAF MIMETES

Unfortunately this plant has become extremely rare in nature and as yet it has not been grown by nurseries. It is to be hoped, however, that plants will become available. It should be watered well and, if tried in hot inland districts, it should be shaded for the hottest hours of the day. The leaves, particularly those near the top of the plant are silver, and glisten in the sunlight. At the top they spread out and allow one a sight of the flowers hidden between them. The flowers of rose and yellow show up beautifully against the silver of the leaves. The plant grows to 1.25 m (4 ft) and is upright in habit.

M. cucullatus ROOISTOMPIE
(*M. lyrigera*)

In early to late spring these gay bushes make a glorious show of colour. The leaves of the plant overlap all the way up the stem. Each leaf measures about 4 cm in length and 12 mm in breadth. Towards the top the leaves change colour in spring and become yellow at the base shading to bright red along the upper half. The flowers cuddled between these topmost leaves add to the colourful effect for they have hairy silvery stamens and prominent red styles. The plant grows to 1-1.5 m (3-5 ft) in height and is upright in habit.

M. hirtus RED-AND-YELLOW BOTTLE BRUSH

This species grows to 1 m (3 ft) and has oval grey-green leaves neatly arranged along the stem. In spring when in flower, it is a fine sight, for the flowers of yellow and red show up between the leaves at the top of the stem, and above them is a showy rosette of rose-pink leaves.

MUNDIA SPINOSA TORTOISE BERRY, DUINEBESSIE

DISTRIBUTION: This pretty shrub can be found growing in fairly dry areas of the north-western Cape and along the coast, from Cape Town to East London.

DESCRIPTION: The plant grows to about 1.25 m (4 ft) with a spread of as much or more, and has graceful, slender, arching stems which end in a sharp spine. The leaves are very small. In dry areas they are rather like those of an erica, but along the coast they tend to be a little broader. In winter the shrub becomes smothered with little flowers which are usually a delicate and very pretty shade of mauve, but occasionally plants with white ones can be found. The flowers, which resemble those of the polygala, are only 8 mm across but they are carried in such profusion that the shrub is a charming sight when in flower. After the flowers fade the plant becomes festooned with scarlet berries which are edible and relished by birds, tortoises and children. At one time Malay hawkers carried these around with other fruit for sale in the streets of Cape Town. An infusion made from the stems was used by early colonists in the treatment of sleeplessness and hysteria and also for flatulence and colic.

CULTURE: This plant does well in areas with a low rainfall and in sandy or gravelly soil. When grown inland it should be watered in autumn and winter to encourage good flowering. More use could be made of it to check the drift of sand at the coast and inland.

Carnival Bush (*Ochna atropurpurea*) is a decorative shrub which does best in warm gardens.

NEBELIA PALEACEA
(*Brunia paleacea*)
DISTRIBUTION: Can be found on mountain slopes and plains from Cape Town to Riversdale.

DESCRIPTION: This shrub which is a member of the brunia family grows to 45-60 cm in height and spread. Its leaves are minute and closely pressed against the stems. In late spring the round flower-heads which measure 5-10 mm across appear in clusters. They are creamy-white in colour and tufty in appearance when mature. This is a dainty shrub.
CULTURE: This and other species of nebelia are worth growing in gardens, but as yet stock is not available from nurseries and little is known of their tolerance of frost. They would need to be watered well during autumn and winter to encourage good flowering in spring.

NYMANIA CAPENSIS CHINESE LANTERN, KLAPPERBOS
DISTRIBUTION: Grows in dry areas of the Karoo, South West Africa and Namaqualand.
DESCRIPTION: This is an excellent plant for dry gardens where winters are cold. The plant grows to 2 m (6 ft) and is upright in habit. The flowers are not showy, but they are followed by large decorative balloon-like seed-pods which are 4 cm across and look rather like paper Chinese lanterns. They vary in colour from green to dusty pink and old rose and are at their prettiest in late winter and spring. The flowers which precede them are brick-red to rose. The leaves, which are small and leathery, emerge from the branches in little tufts.
CULTURE: Nymania grows readily in poor soil but it is advisable to improve the soil to promote quicker growth. As it is tolerant of both drought and cold this is an ideal plant for gardens where the rainfall is sparse and where low temperatures prevail in winter. It grows well in alkaline soil. Nymania is slow-growing, particularly if it is not watered fairly regularly when young. It can be grown from seed or cuttings.

OCHNA ATROPURPUREA CARNIVAL BUSH, OCHNA
DISTRIBUTION: Grows naturally from Caledon, east into Natal and in the warmer parts of the Transvaal.
DESCRIPTION: This shrub grows to 2-3 m (6-10 ft) and is attractive for a long period because of its neat, glossy leaves, which, when they emerge in early spring, are tinged with pink and bronze.

They measure 4 cm in length and have neatly serrated edges. The plant can be kept smaller in size by trimming it back in late summer. It makes a decorative hedge and it is also a fine plant to have as a background to a flower border or interplanted with other shrubs. The flowers, which appear in spring, have five yellow petals, and, after the flowers fade, the plant continues to be colourful, as the sepals become bright crimson, and the large round seeds turn a glistening black and show up beautifully against the red of the sepals.

CULTURE: Ochna grows very readily once it has rooted but it is not easy to propagate from seed or cuttings. Plant it in good soil and water it regularly throughout the year to promote speedier growth. It stands moderate frost but does best in warm, frost-free gardens.

OLDENBURGIA ARBUSCULA KREUPELBOOM

DISTRIBUTION: Grows wild in the eastern Cape.

DESCRIPTION: This is an unusual plant which reminds one of a gigantic specimen of bonsai. It grows to 2-3 m (6-10 ft). The rough bark which clothes the stem gives it a rather gnarled and rugged appearance which accounts for the common name of *kreupel,* meaning lame. The leaves are huge, oval and rounded at the ends. When young they are heavily felted but later become

Oncoba *(Oncoba spinosa)* is suitable for large gardens and parks.

Nebelia paleacea produces dainty and unusual heads of flowers in spring.

The Chinese Lantern or Klapperbos *(Nymania capensis)* stands drought and cold.

The growth of Kreupelboom *(Oldenburgia arbuscula)* is like a gigantic piece of bonsai.

dark green, so that the plant appears to have both green and silver leaves. They emerge in whorls and are 30 cm (1 ft) or more long and 15 cm broad, deeply veined and corrugated. The flowers are carried in large clusters. Each flowerhead looks like a large domed thistle coloured silver-grey and dusky maroon. This is a good accent plant for the rock-garden or for a hot part of the garden.

CULTURE: It grows slowly and stands only mild frost. Plant it in good soil and water well to promote quicker growth.

ONCOBA SPINOSA AFRICAN DOG ROSE

DISTRIBUTION: Grows naturally in Natal and the Transvaal, and further north.

DESCRIPTION: This is a large shrub growing to 3 m (10 ft) with glossy, oval leaves with sharply serrated margins. The sweetly-scented, white flowers are rather striking. They measure 5-8 cm across and have prominent yellow stamens. They look somewhat like a single dog rose. They appear in summer and are followed by large round fruits which turn creamy-yellow when mature. The stems are armed with sharp spines and the plant is useful for making a hedge in a large garden. Sometimes the leaves turn deep bronze in winter and spring. There is a smaller species—*O. kraussiana*—which grows to 2 m (6 ft). This species was at one time popular as a pot plant in England. Its flowers have a more pronounced perfume and the leaf margins are smooth.

CULTURE: Oncoba does best in gardens where winters are mild. To encourage quick growth, water the plants well throughout the year.

PARANOMUS PERDEBOS

DISTRIBUTION: The natural habitat of these little shrubs, which belong to the protea family, is from the Cape Peninsula to Humansdorp.

DESCRIPTION: There are two species which are decorative in the garden and which also provide pretty stems for flower arrangements. They have oval as well as feathery leaves and spikes of flowers at the ends of the stems. The flowers are shaded from mauve and pink to dull rose and green.

CULTURE: Plant them in soil rich in humus and water them regularly during autumn to spring, when they make most growth and flower. They will tolerate sharp frost when well grown, but should have protection from frost when young. They need acid soil.

P. reflexus GREEN PARANOMUS

This species grows to 1.25 m (4 ft) and has feathery foliage along the lower branches and different leaves near the top. The leaves just below the flowerhead are oval and pointed. The flowers of yellowish-green are carried in conical heads at the ends of the stems in late autumn and winter. They consist of slender tubes arranged in a reflexed fashion.

P. spicatus PERDEBOS

This species should be kept moist from autumn to spring but can remain quite dry during summer. It grows to 1 m (3 ft) and bears stems of feathery leaves topped by cylindrical arrangements of little flowers of mauvy-pink or grey and rose. It is a pretty plant when the flowers open and very useful for arrangements as the flowers last well when cut.

PAVETTA — CHRISTMAS BUSH

DISTRIBUTION: The most decorative species occur in the eastern Cape and up through the Transkei into Natal.

DESCRIPTION: The species suitable for the garden are evergreen shrubs growing from 1-3 m (3-10 ft), attractive in form and with pleasing, glossy foliage of dark green. The flowers which are carried in clusters 8-10 cm across are fairly small, white and sweetly scented, with prominent ivory styles which add to the dainty appearance of the flowers. The flowering time is summer.

CULTURE: These shrubs do not do well in gardens where winters are cold but they are fine plants for warm coastal gardens. Plant them in holes in which there is plenty of compost. They grow well in partial shade.

P. lanceolata

Grows to a height of 3 m (10 ft) but can be kept trimmed back to smaller size. In summer it bears an abundance of sweet-scented flowers which show up well against the dark green of the leaves. In its natural habitat it flowers at Christmas.

P. revoluta

(*P. obovata*)

Is a neat shrub to 1 m (3 ft) with rounded leaves which are glossy and leathery. The sweetly-scented, white flowers appear in summer.

PHYGELIUS CAPENSIS — RIVER BELLS

DISTRIBUTION: Is widely distributed from the eastern Cape through Natal and the Orange Free State to the Transvaal, often on hills or lower mountain slopes.

DESCRIPTION: This is a quick-growing plant to

River Bells *(Phygelius capensis).* A quick-growing plant which likes some shade.

1 m (3 ft). It sends up numerous stems from the ground clothed with soft leaves broad at the base and pointed at the apex, with finely serrated margins. The flowers which come off the upper part of the stems are tubular in form and salmon to coral-red in colour shaded with yellow inside. They hang gracefully from the stems. *P. equalis* has flowers of rose-pink. They flower in late spring and early summer.

CULTURE: Plant it in good soil and water regularly. In hot inland districts it does better if planted where it is shaded during the hottest part of the day. It may be cut down by frost but the plant grows up quickly again and flowers once more by summer. As the flower stems tend to keel over it is advisable to stake the plants.

PHYLICA PUBESCENS — FEATHERHEAD

DISTRIBUTION: Is limited in habitat to the south-western Cape where it is usually found on lower mountain slopes.

DESCRIPTION: This shrub is unusual in appearance and provides fine material for arrangements. The whole nature of growth of the plant is decorative for it bears graceful stems wreathed in leaves. They are slender, hairy and of subtle shade of greyish-green to yellowish-green, and they glisten in the sunshine. The tops of the stems end in a whorl or rosette, with a sparkling feathery appearance. It is at its best in late winter and early spring. The plant grows to 1.25 m (4 ft).

CULTURE: Plant it in good soil to which some compost or leaf mould has been added, and water it regularly from autumn to spring. It is quick-growing but not hardy to severe frost.

PLECTRANTHUS SPUR FLOWER

DISTRIBUTION: Grow in shady places in forest areas from the eastern Cape up into Natal and the eastern Transvaal.

DESCRIPTION: Plectranthus include quick-growing shrubs and perennials which are invaluable for bringing colour to the garden in late summer and early autumn. The shrubby types described below have fairly large leaves and showy spikes of flowers of pink, mauve or a glowing purple. Each flower consists of a short or long tube ending in a two-lipped face.

CULTURE: Plectranthus grow very quickly and easily, but they should be watered regularly during dry periods and particularly during summer. In inland gardens they should be planted in a shady place, but in coastal gardens they do not mind being in full sunshine. To promote fine flowering, plant them in good soil. They stand moderate frost. Most species make good tub plants for a shady part of the patio or garden. If the plants tend to grow too large for their allotted space, cut them back in late winter or early spring. They grow very readily from cuttings.

P. behrii PINK SPUR FLOWER

This quick-growing shrub occurs in the Eastern Cape and the Transkei and is worth trying in gardens where frosts are not extreme. It grows to 60-90 cm (2-3 ft) and has rose-coloured stems with quadrilateral and grooved sides. The leaves are large, rough and heart-shaped, green on the upper surface and purple on the lower, with serrated margins. The flowers which are carried in showy spikes are pale pink to rose. The flowering time is late summer and autumn. In hot, dry gardens plant it where it is shaded by other plants or a building during the hottest hours of the day. This is a handsome plant for a shady part of the garden or patio.

P. ecklonii PURPLE SPUR FLOWER

Grows to 1.25-2 m (4-6 ft) in height and spread. The stems are four-sided and the leaves are large, rough in texture, and have serrated margins. They have prominent wine-red veins on the underside. The spikes of mauve to purple flowers are extremely attractive and appear in autumn when there is not much else in flower in the garden. It looks effective growing next to Port St. Johns climber, plumbago or yellow Cape honeysuckle, all of which flower at the same time. It does best in a shady place.

Pink Spur Flower *(Plectranthus behrii)*.

Purple Spur Flower *(Plectranthus ecklonii)*.

Featherhead *(Phylica pubescens)*.

Plumbago *(Plumbago auriculata)*.

PLUMBAGO AURICULATA PLUMBAGO, CAPE LEADWORT
(P. capensis)

DISTRIBUTION: Commonly found in the eastern Cape from Humansdorp up through the Transkei to Natal.

DESCRIPTION: This cheerful shrub brings life to the bare veld even during periods of prolonged drought. It is a scrambling shrub but can be trimmed to form a neat hedge or specimen shrub of distinct shape. It is only too ready to grow and may have to be kept within bounds or it will take up too much space in the garden. The plant is clothed in masses of small slender leaves which are pretty throughout the year, and in summer it becomes covered with heads of sky-blue flowers for many weeks. It can be had with white flowers, too, but these are not as decorative as the blue. Its main flowering period is summer, but when grown in warm areas it flowers off and on during other seasons of the year, too. Each flower has five rounded petals and measures 18 mm across. This is an excellent shrub to use as a background plant, as a specimen shrub planted in a tub and trimmed to shape, as a hedge or as a windbreak of moderate height.

In the book *Wild Flowers of the Eastern Cape Province* it is mentioned that the Xhosa people make use of it medicinally and in magic. The powdered root is used to relieve headaches or wounds. A piece of the stem placed in the thatch above the door is said to ward off lightning, and the outer part of the root added to the water they wash in is thought to be efficacious in bringing together an estranged husband and wife!

CULTURE: This is one of the easiest of shrubs to grow from root sections or cuttings. It does not demand good soil and once it is well grown it will stand long periods without water. Unfortunately it is cut down by severe frost but this is no reason for not trying it in gardens where winters are severe, as very often it will grow up again quickly in spring and be full of flowers by summer.

PODALYRIA SWEETPEA BUSH, KEURTJIE

DISTRIBUTION: Grows naturally from Cape Town to the eastern Cape.

DESCRIPTION: These are attractive small and large shrubs which produce sweet-scented, pea-shaped flowers mainly mauve or pink, and sometimes white. They flower in late winter and spring.

CULTURE: They are quick-growing and are not particular as to soil. They should be watered fairly regularly to promote good flowering but they will stand long periods without water when mature. In gardens which have severe frost they should be grown in a protected position as they are not hardy to more than moderate frost.

P. burchellii

This is an attractive plant from the eastern Cape.

The shining silver leaves of the Silver Sweetpea Bush *(Podalyria sericea)* make a fine foreground to Cyclopia.

The sweetly-scented flowers of Keurtjie *(Podalyria calyptrata)* appear in late winter.

Zimbabwe Creeper *(Podranea brycei)* is a vigorous and quick-growing climber which flowers in summer and autumn.

It grows to 45 cm ($1\frac{1}{2}$ ft) and has glistening, silvery leaves which are highly decorative. In late winter it becomes covered with cyclamen-pink flowers which show up beautifully against the grey of the leaves. This is a desirable plant for small gardens where space is a limiting factor.

P. calyptrata KEURTJIE

This tall bush or small tree grows on hill slopes near Cape Town and east towards George. It reaches a height of 3-4 m (10-13 ft) and has oval leaves 4 cm long and fairly large, very sweet-smelling, pea-like flowers usually mauve or pinkish-mauve, and occasionally white. It can be used in the garden as a small tree or it can be trimmed after flowering to keep it to shrub size. It is quick-growing and fairly hardy to frost and a delightful sight in August and September when it is in full flower.

P. sericea SILVER SWEETPEA BUSH

This is a most attractive shrub from the eastern and south-western Cape. It grows to 1 m (3 ft) in height and spread, and has small oval leaves covered with silvery hairs which make the whole plant glisten in the sunshine. The flowers are small, generally mauve, but sometimes white, and sweetly scented. When they fade in spring the plant becomes covered with large, attractive, silver pods which are even prettier than the flowers, and these persist into summer. When grown in gardens which have frost this plant should be protected. It should be grown as a background plant in a bed of flowers or in front of a shrubbery, and it looks splendid in a tub, particularly when it is near flowers of yellow, blue or pink, which show up well against its silver leaves.

PODRANEA RICASOLIANA PORT ST. JOHN'S CLIMBER

DISTRIBUTION: Is found in a restricted area near Port St. John's on the Pondoland coast.

DESCRIPTION: Although we have many attractive native shrubs and flowers there are few indigenous climbers of great merit. This one, however, is a lovely climber worth a prominent place in the garden. The leaves are divided into attractive oval, pointed leaflets of glossy green. The stems become woody as they age and support the top growth to some extent but the plant should have some support to keep it from sprawling too much. The beautiful flowers of pale pink appear in handsome clusters in summer and early autumn. This plant is very similar to the Zimbabwe creeper *(P. brycei)*. The differences are that the latter has narrower and lighter leaves and smaller flowers of a deeper colour.

CULTURE: The same kind of growing conditions suit these two plants. Plant them in holes filled with compost and good soil and water them regularly during spring and summer. They do not mind being dry during winter. If the plants grow too large prune them back in late winter or early spring. Once established they stand moderate frost and quite considerable drought. They can be raised easily from cuttings and are quick-growing.

POLYGALA POLYGALA, PURPLE BROOM, BLOUKAPPIES

DISTRIBUTION: Several species are to be found in the eastern Cape, the western Cape, Natal and the Transvaal.

DESCRIPTION: They have flowers from mauve to purple and sometimes white. The flowers usually appear in late winter and early spring and look

Polygala *(Polygala myrtifolia)* produces its charming flowers in spring.

somewhat like those of a pea, each one having a keel from which emerge tufts of silky hairs.

CULTURE: The species of most decorative value in the garden grow readily and stand moderate frost. Even when frosted they generally recover. They are quick-growing plants which do best if watered regularly during autumn and winter.

P. myrtifolia SEPTEMBERBOSSIE
Grows to 1.5 m (5 ft) with a spread of as much and has oval, pointed, light-green leaves about 3 cm long, and little clusters of unusual flowers of a pleasing shade of purple. The wings of the flowers are darker than the keel, which is attractively veined when young with mauve and green. From the keel there emerges a little cockade of lilac-coloured silky hairs. It flowers early in spring.

P. virgata PURPLE BROOM
This tall, erect-growing shrub to 2 m (6 ft) looks somewhat like the yellow broom in form. It has slender leaves 2-4 cm long and less than 6 mm wide. They are carried sparsely on the round, rush-like stems. In early spring its decorative purple flowers appear in long graceful spikes. The shrub is colourful for many weeks.

PRIESTLEYA VILLOSA SILVER PEA
(P. tomentosa)
DISTRIBUTION: This shrub has become rare and is now seldom found in its native haunts on the slopes of mountains in the Cape Peninsula.

DESCRIPTION: It grows to 1.25 m (4 ft) or more and has small oval silver leaves which glisten in the sunlight. The rounded heads of yellow, pea-shaped flowers appear at the ends of stems from autumn through to spring. *P. hirsuta* is a rangy plant with similar flowers but less pleasing foliage.

CULTURE: This plant can stand moderate frost but not long periods without water. It should be planted in soil to which plenty of compost or leaf mould has been added, and watered well and regularly during autumn to spring.

PROTEA VARIOUS COMMON NAMES
DISTRIBUTION: Although most of the hundred or more species of these unique and often handsome plants are native to the south-western Cape, there are several species to be found further east and north.

DESCRIPTION: It is only within the last decade that proteas have become popular as garden subjects in South Africa and they are still not being planted as widely as they should be. It is interesting to note that they were grown in England and Holland from early in the seventeenth century. Some species have been grown commercially in New Zealand to produce flowers for florists, for

Purple Broom *(Polygala virgata)* with its richly coloured flowers makes a fine show for a long time.

several years, and they were being planted by gardeners in New Zealand and Australia before they became generally known in their homeland.

Proteas are not difficult to grow, and, being fairly adaptable they will flourish in many parts of the country provided some attention is paid to their likes and dislikes. This is something one has to do with many other plants, such as azaleas, in order to have good results. The little attention they demand is more than repaid in the number of handsome flowers they produce. There are two other good reasons for growing proteas. One is that many of them flower in autumn and winter when the garden may be rather bare, and the second is that they produce flowers which last for a long time on the plants and in arrangements. They are at first not easy flowers to arrange but, with a little practice, one begins to know just how to display them to best advantage. An arrangement of proteas retains a fresh appearance for a much longer time than arrangements of other plant material. Some species can be used effectively in dry arrangements, too. Until now the proteas most generally grown have been those with large flowers, but the species with small flowers are in some cases more suitable for town gardens and they also produce flowers which are easier to use in small arrangements and corsages. The following are the names of some with comparatively small flowers: *Protea cedromontana, P. lacticolor, P. minor, P. nana, P. odorata, P. pendula, P. pityphylla, P. scolymocephala* and *P. sulphurea*.

Where to plant Proteas: However small the garden there is space for one or more proteas, and in large gardens a great deal of space could be allocated to them. Generally they look best when planted in groups of two or three with foliage which is more or less the same. It is a good idea to plant them near to one another not only to produce an effective show, but also to make it easier to provide them with the right soil and growing conditions. Grown in groups they can make the frame to the garden, or they can be planted along a drive, or as part of a shrubbery. The small species look effective in a rock-garden.

The proteas listed below are grouped according to size to enable the gardener to choose species to fit in with the size of the garden, or to plant them so that the shorter ones are not hidden by taller ones.

PROTEAS LARGE IN GROWTH

These grow to 2-3 m (6-10 ft) or more with a fairly wide spread. They are better suited to large gardens and parks than to the garden of average size in town. They can be used as an informal hedge, as screen plants or as specimen shrubs.

P. arborea
P. caffra
P. compacta
P. eximia
P. lacticolor
P. laurifolia
P. longiflora
P. macrocephala
P. mundii
P. neriifolia
P. obtusifolia
P. punctata
P. repens
P. rouppelliae
P. rubropilosa
P. susannae

MEDIUM AND SMALL IN GROWTH

These are suited to small as well as large gardens. They are less than 2 m (6 ft) in height and spread.

P. acaulis
P. amplexicaulis
P. aristata
P. barbigera
P. cedromontana
P. cynaroides
P. effusa
P. grandiceps
P. harmeri
P. lepidocarpodendron
P. longifolia
P. lorifolia
P. minor
P. nana
P. pendula
P. pulchra
P. rupicola
P. scolymocephala
P. speciosa
P. stokoei
P. sulphurea
P. witzenbergiana

FOLIAGE

When making a choice of proteas for the garden the nature of the foliage of the plants should be considered, as flowers are ephemeral and highlight the garden only for a fairly short period whilst leaves remain effective throughout the year. Some proteas have leaves with a greyish tinge which look effective as a contrast to the greens of the leaves of other plants. Many have fairly large leaves, sometimes broad and rounded and sometimes long and slender. Others have leaves which are almost needle-like and therefore more graceful in appearance. The following are the names of some of the proteas with slender needle-like leaves: *Protea acerosa, P. aristata, P. nana* and *P. pityphylla*. Before choosing proteas for the garden take into account the form and colour of the foliage of the different species.

FLOWERS

The protea has flowers densely packed together in heads which may be large, of moderate size or fairly small in size. The flowers are slender tubes closely packed together to form a head surrounded by colourful and decorative bracts, which in some cases enclose the flowers tightly whilst in other cases they spread out and look like a ray of petals. The central part of the flowerhead is often attractively coloured, and the bracts are usually very handsome and of delightful shades, varying from soft silvery pink to vivid rose and crimson, from cream to yellow and from pale cyclamen to burgundy or cinnamon, often tipped with attractive fringes of silky hairs. Very often too, the bracts have a rich sheen and look as though they have been fashioned from wax, glass or velvet.

Because it is not easy to choose plants for the garden unless one knows them well I add here a list of ten proteas from which to make a selection. Those with large gardens may have space for many more species than the ten listed:

Protea aristata (CHRISTMAS OR LADISMITH PROTEA)
P. barbigera (BEARDED OR QUEEN PROTEA)
P. compacta (BOT RIVER PROTEA)
P. cynaroides (KING OR GIANT PROTEA)
P. grandiceps (PEACH PROTEA)
P. neriifolia (OLEANDER-LEAFED PROTEA)
P. pulchra (GLEAMING PROTEA)
P. repens (SUGARBUSH PROTEA)
P. scolymocephala (SMALL GREEN PROTEA)
P. speciosa (BROWN BEARDED PROTEA)

CULTURE

Most of the proteas grow on hillsides and mountain slopes of the south-western Cape, where the rainfall is regular from autumn through winter to mid-spring. Although the summers in this part of the country are hot and dry, the areas where most of the proteas grow are swept by south-east winds in summer which bring some moisture to mountain slopes, or they are near enough to the sea to benefit from coastal humidity.

The soil where most of them grow is acid and well-drained and to ensure success in growing them in the garden, it is important to see that the soil is acid, except for the few species which do not seem to mind an alkaline soil. In gardens where soil is alkaline it is advisable to dig holes 60 cm (2 ft) deep and across, and to fill these with acid compost or leaf mould mixed with a little of the soil previously removed from the hole. The leaves of oaks and pines are particularly useful in providing acid leaf mould. In areas where the water tends to reduce the acidity of the planting mixture apply a sprinkling of aluminium sulphate (alum) or sulphur to the soil once a month, to ensure that the soil does not become too alkaline for the plants.

Most proteas seem to do best where the pH of the soil is not higher than 5.5. Particulars of the significance of the term pH are given on page 11. Generally it can be said that where hydrangeas are distinctly blue it can be assumed that the soil is acid. Growers on a large scale should, however, take the precaution of having their soil tested to ascertain its nature in this respect. In areas where it is difficult to ensure that the soil mixture remains acid the following species should be tried as they appear to be more tolerant of alkaline conditions: *Protea aristata, P. barbigera, P. cedromontana, P. cynaroides, P. eximia, P. lacticolor, P. laurifolia, P. longiflora, P. macrocephala, P. minor, P. neriifolia, P. obtusifolia, P. repens, P. susannae.*

It has been suggested that proteas do best in soils which drain easily because the roots secrete a substance which is poisonous to them if it is not leached away. Research is at present being done to establish whether, in fact, this is an idiosyncracy of the protea. If the soil in your garden is a heavy clay which does not allow water to drain through easily it is advisable to make holes 1 m (3 ft) deep and across and to incorporate plenty of rubble in the bottom of the holes so that no water stands about the roots of the plants for a long

The Sprawling Protea *(Protea amplexicaulis)* carries its flowers close to the ground usually facing down.

The new leaves of Waboom *(Protea arborea)* are almost as decorative as the flowers.

The Cedarberg Protea *(Protea cedromontana)* has dainty flowers of unusual colouring.

time. This is necessary only in areas of heavy rainfall where the ground is not on a slope.

When proteas are grown in regions other than the winter-rainfall region, they should be watered regularly from autumn to spring, which is their period of greatest growth and the flowering period of a large number of them. A thorough soaking once or twice a week is what is required and not a sprinkling of the surface of the soil. A thick mulch of straw or old leaves on the ground around the plants will help to promote good growth.

When once they are established many species of protea will stand the rather sharp frosts experienced in some inland gardens but they should be protected for the first year or two. This is easily done by inverting a large cardboard carton over each plant during the night, but it should be taken off in the daytime as they like to have a free circulation of air about them. For this reason it is inadvisable to plant them against the walls of a building or hemmed in by large shrubs. They can however be planted where they will be shaded by a tree from the early morning winter sun until about 10 a.m., as this will also help to prevent frost-damage to leaves and flowers. As many proteas have fine roots near the surface as well as deeper-growing ones, it is inadvisable to cultivate the ground around them unless it becomes unduly hard, when the crust can be loosened with superficial cultivation, avoiding disturbance to the surface roots, if possible.

Cutting Back: Proteas do not drop their heads, as other plants do, when the flowers have faded. The faded flowers remain on the protea plants, but close up to protect the seeds which do not mature for several months after the flower has faded. To keep the plants tidy and shapely it is advisable therefore to cut off the old heads with a good length of stem. This can be done immediately after the flowerhead has faded unless seed is required, in which case it must be postponed for several months until the seed is ripe. The cutting off of faded flowers with several inches of stem helps to keep the plants neat and encourages fresh basal growth.

Planting Time: The best time to set out young plants in the Western Province is in early autumn when the rains start, and the best time for planting in the summer-rainfall region is from September to November, i.e. at the beginning of the rainy season. Gardeners will find that many nurseries have proteas for sale, and buying plants will ensure having flowers within a year or two, whereas growing plants from seed necessitates waiting three to four years for flowering plants to develop. Many nurseries now send out protea plants in plastic containers which can be removed easily without disturbing the soil about the roots but, if they are sent in tins, care must be taken to get the plant out of the tin with the soil intact about the roots. Before transplanting, water the soil in the tin well, then squeeze the top of the tin all around and give the bottom a few sharp taps with a trowel. This will invariably make it easier to remove the plant with the soil still clinging to the roots.

Plants purchased from a nursery may be quite small. One-year old plants often transplant and progress better than larger ones, and for this reason most nurseries prefer to send out one-year old plants. Although many proteas will continue to grow for years some of the most decorative species tend to become woody after a few years and to produce fewer flowers. It is therefore a good idea to take out such plants after five or six years of flowering and to replace them with young plants.

Growing Proteas from Seed: Sow the seeds in late summer or early autumn (March-April) either in beds specially reserved for growing seedlings, or in pots sunk into the ground which can later be removed. They should be sown where they are exposed to sunlight and not in a heavily shaded place. The soil should be well mixed with acid compost and leaf mould, and watered regularly throughout the year, and particularly until the plants have germinated. The drainage in the bed should be good so that water can leach away the toxins thought to be present and capable of inhibiting germination. When sowing them in pots it is important to see that the soil in the pot will allow the water to drain through the bottom of the pot quickly. Experiments have shown that soaking the seed does not produce as good germination as sowing the seed in a soil medium which allows for good drainage, and keeping it moist. It is inadvisable to transplant seedlings to the garden until they are a year or more old.

P. acaulis DWARF GREEN PROTEA

Although most protea plants are fairly large in size there are some species which grow close to the ground and produce flowers just above the ground from their underground stems. This

The Christmas or Ladismith Protea *(Protea aristata)* produces its lovely flowers later in the year than other species. ▷

species bears its bowl-shaped flowers at ground level. Each flowerhead measures about 5 cm across and has bracts of pale lime-green often edged with crimson. Occasionally one with red bracts can be found.

P. acerosa
This species can be found growing in sandy soil fairly near the sea between Cape Town and Hermanus. Its slender stems with short needle-like leaves grow to a height of about 30 cm. The globular heads of flowers are small, measuring only 4 cm across. They are carried tightly clustered together on the ground. This species with its cinnamon to tan flowers would make an interesting plant for the rock-garden.

P. amplexicaulis SPRAWLING PROTEA
This is a sprawling species which looks attractive growing over a bank or wall. Its leaves have a reddish tinge when young, and the stems with their leaves make a carpet along the ground. The flowers seldom make a show in the garden for they tend to be hidden by the leaves but they are effective in arrangements. They measure 7 cm across and the bracts surrounding the flowers are short, forming a bowl about the central mass of flowers. The colour is chocolate-maroon on the outside and cinnamon to biscuit-yellow inside. This species stands a good deal of frost and slightly alkaline soil. It flowers in winter and early spring.

P. arborea WABOOM
(*P. grandiflora*)
This protea was used by the early settlers for making the wheels of wagons. It grows to about 3-4 m (10-12 ft), and higher under optimum conditions, and has a rather attractive, gaunt form. It is a good plant for large gardens where it should be planted against a background of dark green foliage to show up its handsome grey-green leaves. Formerly the leaves and bark were used for tanning, and thousands of trees were destroyed because of this. It is said that the Voortrekker leader Louis Trichardt used ink made from Waboom leaves to write his diary. The flowerheads which appear in late autumn and winter are attractive. They are about 10 cm across, rather like a large round shaving brush, and of unusual shades of cream to lime-yellow. There is also a form with flowers of deep rose. In spring the new leaves grow in whorls, which look like flowers from a distance as they are prettily coloured from rose-mahogany to lime-green.

The tufted flowers of Waboom *(Protea arborea)* make an attractive show.

Bearded Protea *(Protea barbigera)*. A large and handsome species.

P. aristata CHRISTMAS PROTEA, LADISMITH PROTEA

This lovely protea has only recently been introduced into cultivation. It grows to 1.5 m (5 ft) and its long, needle-like leaves clothe the stems right up to the flower, giving the stems a feathery appearance. Both the leaves and the flowers of this species are most attractive. The flower itself is about 10-12 cm in length and coloured dusty pink to a rich shade of rose. It flowers in spring and summer. This species stands quite considerable frost but no dryness in winter.

P. aspera

This is another ground protea. It has long, slender leaves which encircle the flowers. The flowerheads are small with tan hairs at the centre and bracts of biscuit-yellow and cinnamon-brown. It grows in the Caledon district fairly near the coast. The flowering time is spring to early summer.

P. barbigera BEARDED PROTEA, GIANT WOOLLY PROTEA, QUEEN PROTEA

Is a magnificent species which grows to 1.5 m (5 ft) with a spread of almost as much. The leaves are 15-20 cm in length and 2-3 cm broad and of a pleasing shade of grey-green with margins of rose. The flowerheads are huge—often as much as 15-20 cm (6-8 in) across. The inside of the flowerhead is silvery-white ending with a central cone of gleaming black. The bracts, which are fringed with silky white hairs, are of a luminous deep pink, rose, palest yellow or lime-green. This protea stands fairly severe frost. It flowers in late winter and spring.

P. caffra HIGHVELD PROTEA

Can be found growing naturally near Johannesburg and elsewhere in the Transvaal, and on mountain slopes in Natal and Lesotho. It is a somewhat rangy shrub or tree growing to about 5 m (16 ft). The trunk has a gnarled appearance which is rather pleasing and the grey-green leaves are about 15 cm long and lance-shaped. The new growth which is rosy-pink is very attractive in spring. The flowerheads are most profuse in summer and quite attractive. They are 6-8 cm across, with rose-pink bracts spread out like rays around the central cone of pink to white flowers.

P. cedromontana CEDARBERG PROTEA

A small protea growing to 1 m (3 ft) or more. It has long slender leaves 12 cm long and 12 mm broad. The flowerheads are less than 5 cm across and of unusual colouring, for the central mass of flowers is burgundy and cinnamon and the surrounding bracts are brick-red to crimson. The topmost leaves stand up around the flowerheads partially hiding them from view. It stands sharp frosts. The flowering time is late winter and early spring.

P. compacta BOT RIVER PROTEA

This is one of the most rewarding of the proteas as it flowers from winter to mid-spring. It is a tall shrub of erect form to 3 m (10 ft). The flowerheads are 10 cm in length and carried proudly erect. They are of the most enchanting shades of rose-pink and look as though they have been fashioned from velvet. An ivory-coloured one is also known. This species blooms profusely for a long time. The leaves are oval and light green, neatly overlapping one another along the stems. This is one of the most decorative for flower arrangements in winter and early spring.

P. convexa

This unusual protea can be found in mountain regions bordering the Karoo. Because the flower emerges at ground level it is not highly ornamental except in a rock-garden. The flowerhead is, however, very attractive. It is bowl-shaped with a silvery sheen to the central dome of flowers, and has bracts which are lemon-yellow on the inside and dusty-rose on the outside. The whole flowerhead measures about 10 cm across. The leaves of this species are large and broad. It flowers in late winter.

The Bot River Protea *(Protea compacta)* is decorative.

P. cryophila SNOW PROTEA

This species is known as the snow protea not only because of its appearance but also because its natural habitat is the Cedarberg Mountains where snow is not uncommon in winter. Because of its habit of growth it should be regarded as a curiosity rather than a plant of ornamental value in the garden. The plants spread their long leaves across an area of about 3 m. Each leaf is 45-60 cm in length and folded along its length. The flowerheads which appear in mid-summer measure about 12 cm in length and consist of closely packed, slender white woolly bracts surrounding the flowers.

P. cynaroides GIANT OR KING PROTEA

This is a protea of rounded form growing to about 1 m (3 ft). It has leaves which vary somewhat in shape but they are usually leathery and carried on fairly long leaf-stalks. The flowerheads appear at the ends of stems and are quite outstanding as they measure up to 25 cm (10 in) across. The heads vary somewhat in shape and colour according to habitat. Generally the central mass of flowers is ivory to silvery-white, pale pink or pale lime-green, and the bracts which open out wide are of the most entrancing shades of pink to rose, with a velvety texture. A single flower makes a spectacular show for a long time. The plant needs protection from frost when young and is not hardy to severe frost. The main flowering period is from winter to late spring, but the one which grows near Grahamstown flowers in summer too.

P. decurrens

This is another species more suitable for the front of a rock-garden than any other part of the garden. It grows at ground level and produces stems of slender leaves to a height of about 30 cm. The plant grows along the ground and the flowers which are bowl-shaped and only 4 cm wide are generally hidden by the foliage. They are a soft shade of salmon pink. The flowering time is winter.

P. effusa
(P. marlothii)

This protea is not yet offered by nurserymen but one hopes that it soon will be available as it is rare in nature and deserves to be preserved. It grows to 1 m (3 ft) and has leaves which are different from the other two small proteas it most resembles—namely *P. nana* and *P. pityphylla*. They have needle-like leaves whilst *P. effusa* has

Bearded Protea *(Protea barbigera)* showing colour variation. ▷

The Ray-flowered Protea *(Protea eximia)* is unusual in form and colour.

Baby Protea *(Protea lacticolor)*. Its flowers show up well against the grey-green of the leaves.

The Giant or King Protea *(Protea cynaroides)* has large symmetrical flowers. The picture shows one found near Grahamstown which is somewhat different from those which occur in the south-western Cape.

leaves 5 cm or more in length and 12 mm wide. The flowerhead is a shallow bowl 5 cm or a little more across, with satiny bracts of rose-mahogany on the outside and paler on the inside. It flowers in late autumn and winter.

P. eximia RAY-FLOWERED PROTEA
(*P. latifolia*)

Grows to a height of 2 m (6 ft) and is erect in form. The broadly oval grey leaves with crimson margins are neatly arranged around the stems. The flowerhead is beautiful when young but tends to become untidy as it matures. The central mass of flowers is lime-green to rose, tipped with chocolate-brown and fuzzy at the top. The surrounding bracts are spoon-shaped and pale pink to rose, edged with a sparse covering of short silvery hairs at the tip. It stands quite considerable frost and appears to be able to grow quite well in alkaline soil. It flowers from autumn to spring.

P. grandiceps PEACH PROTEA
This handsome protea deserves a place in every garden. It is a neat plant growing to 1-1.5 m (3-5 ft), and is rounded in form with very attractive glaucous (grey-green) leaves edged with rose, 15 cm long and 10 cm broad. The flowerhead is 12 cm long and has beautifully coloured bracts of coral to a rich rose-pink. The upper bracts are edged at the tips with long silky hairs of cinnamon and silver. These bracts bend over the central mass of flowers which are ivory to palest green. It flowers in winter and spring and stands fairly sharp frost.

P. harmeri
Grows on mountain slopes and produces its flowers from autumn to spring. It reaches a height of about 1.25 m (4 ft) and has leaves with a greyish tinge which show up the flowers rather prettily. The flowers are like a shallow bowl with short bracts of rose-red surrounding a central mound of biscuit-yellow flowers.

P. lacticolor BABY PROTEA
A species found in the eastern Cape which grows to 3 m (10 ft), and is suited to the large garden rather than the small. The leaves are oval and leathery and about 5 cm long. The flowerheads measure 5 cm across and have pale pink or cream bracts which spread out wide around the central mass of flowers, which are cream and pink. It stands a good deal of frost and alkaline soil. The main flowering period is autumn to spring but it may bear some flowers in summer too. The species *P. punctata* is very similar to this one.

Peach Protea *(Protea grandiceps)*. The grey of its leaves shows up the fascinating colour of the flowers.

P. lanceolata
Is mentioned here because it is available in the trade and although it is not worth a place in the decorative part of the garden it may be planted by those gardeners who wish to make as complete a collection of proteas as possible. The plants grow to 1.5 m (5 ft) and bear small flowerheads composed of white flowers and bracts. Neither the bracts nor the flowers are closely packed, as they are in many of the proteas, but are rather loosely arranged. The flowering time is winter to early spring.

P. laurifolia FRINGED PROTEA
(*P. marginata*)
This is a tall plant growing to 2.5 m (8 ft) with a wide spread, and is suitable for large gardens and parks rather than for those of average size. It is somewhat similar to *P. neriifolia* but has broader and thicker leaves. The bracts are usually pink with a silvery sheen, edged at the tips with long hairs which are mostly black, but there is some

white amongst them. The flowers rise to a dome which is oval and hairy, and usually yellow in colour. This species flowers in spring. It stands moderate frost and seems to do quite well in alkaline soil.

P. lepidocarpodendron BLACK-BEARDED PROTEA

Reaches a height of 2 m (6 ft) and has green leaves 10 cm long and 2-3 cm wide. The flowerhead is long and fairly narrow and the bracts do not open wide but remain closely erect about the flowers. The bracts are greeny-white tipped with black, with a beard at the top. There are short brownish bracts about the bottom part of the flowerhead. It tolerates alkaline soil and seems to be able to stand more dryness than most species. Its flowering time is winter.

P. longiflora LONG-BUD PROTEA

This is a tall plant more suited to the large garden than the small. It grows to 3 m (10 ft) and has leathery, oval, pointed green leaves. The flower buds, which are slender and about 12 cm long, open wide to display the long silky styles which stand erect. The flowers are elegant but they do not last long, and the faded ones should be removed as they make the plant look untidy. The flowers are rose, pale pink or cream. Its main flowering time is late spring and summer. It should be tried in gardens where the soil is alkaline.

P. longifolia LONG-LEAFED PROTEA

A species growing to 1-1.5 m (3-5 ft) with long, slender leaves 15 cm long by 2 cm wide. The flowerhead is long also—about 12 cm—and composed of bracts coloured ivory suffused with pale green or pale pink. They surround a fluffy mass of hairy, white flowers ending in a pointed tip of black, rather like those of *P. barbigera*. It flowers from autumn to spring.

P. lorifolia

(P. macrophylla)

This species, which grows wild in the Swartberg Mountains, reaches a height of 1.5 m (5 ft) and has greyish-green leaves measuring up to 25 cm in length. The flowerheads, which appear in spring, are made up of a cone of closely packed flowers of pink to rose, with surrounding bracts of the same shades.

P. macrocephala GREEN PROTEA

(P. incompta)

This species has flowers which look exceptional in arrangements. The plant grows to 3 m (10 ft) and is upright in form. The leaves are long and slender and the flowerhead is of moderate size—about 10 cm in length and 5 cm wide—narrowing at the top. The bracts which envelop the flowers never open wide, but they are of a delightful soft shade of green, tipped with white hairs, and they curve over the mass of white flowers in the centre making a neat and handsome flowerhead. There

A pyramidal arrangement which includes: *Protea barbigera*, *Protea compacta*, Leucospermums, Leucadendrons, Berzelia and Erica.

Long-bud Protea *(Protea longiflora).*

Long-leafed Protea *(Protea longifolia).*

are also soft hairs on the lower bracts. It flowers in late autumn and winter. This species appears to be tolerant of alkaline soil.

P. minor GROUND ROSE, AARDROOS

Is a low, spreading plant with charming cup-shaped flowerheads partially enclosed by the long slender leaves which measure about 15 cm long and 6 mm wide. The outer bracts are a waxy dark brown or rose colour and the inner ones are rusty rose on the outside and frame the mass of flowers. These are ivory at the bottom shaded to pink, with a mass of silvery-grey and maroon hairs rising to a peak at the centre. It flowers in late autumn and winter. It grows fairly well in alkaline soil.

Protea mundii

This is a species which grows with exuberance and may finally reach a height of 4 m or more, with a spread of at least half this. It is a rangy shrub suitable only for the large garden or for the collector who wants to gather together as many species as possible. The flowers are similar to those of *P. lacticolor.* They measure 6-8 cm in length and are either pink or white in colour. In this species the bracts do not open as wide as in *P. lacticolor* and there are little rounded heads at the tips of the styles, which are not found in *P. lacticolor.* This species appears to tolerate alkaline soil well, and it flowers later in the year than most of the proteas.

P. nana MOUNTAIN ROSE,
(*P. rosacea*) BERGROOS, SKAAMROOS

Grows to 60-90 cm (2-3 ft) and has arching stems with soft needle-like leaves about 2-3 cm long closely arranged along the upper stems giving them a graceful, feathery appearance. The little flowers hang down hiding their faces but, with the light shining through them, they look most attractive. The flowerheads are 5 cm across and made up of rather short bracts forming a bowl about the mass of flowers inside. The bracts are of an unusual shade of rose-mahogany flushed with green at the base. It bears masses of flowers through winter to spring. Once established it stands fairly sharp frosts.

P. neriifolia OLEANDER-LEAFED PROTEA

This is one of the most handsome of the proteas and it is also the first of which there is a known record in Europe. It was described and illustrated

The drooping heads of the Mountain Rose *(Protea nana)* are attractive.

The Ground Rose *(Protea minor)* is a good plant for the rock-garden.

by Clusius in a work published in Antwerp in 1605—half a century before Van Riebeeck came to the Cape. The plant grows to 3 m (10 ft) which makes it too large for the small garden unless a single one be used as a specimen shrub. The leaves are long and rather slender, somewhat like those of the oleander, whilst the leaves of *P. laurifolia*, which closely resembles this one, are thicker and broader. It has long flowers 15 cm or more in length made up of beautiful glistening bracts of bright rose or cream tipped with a decorative fringe of dark-brown or black hairs. There is a certain amount of variation in the colour of the bracts. They never open wide and the mass of flowers inside make a hairy dome often tipped with deep maroon, very like those of *P. laurifolia*, except that in the latter the central mass of flowers is usually yellow. The plant flowers well during winter. It tolerates frost and grows quite well in soil which is slightly alkaline.

P. obtusifolia BREDASDORP PROTEA

A large-growing plant to 3 m (10 ft) suitable for gardens where space is not limited. It has leaves 10 cm long and 2-3 cm wide with red margins. The flowerheads resemble those of the sugarbush (*P. repens*), but the leaves are slightly broader and the flowerheads a little shorter. The waxy bracts which overlap one another are rose to crimson or green, tipped with pink to dusky rose. The inner ones are spoon-shaped at the ends. There is also an ivory-coloured form. The inside mass of flowers is silver or pinkish-grey. This is one of the easiest of the proteas to cultivate. It grows in alkaline or acid soil. The flowers appear in autumn and winter.

P. pendula HANGING PROTEA

A small protea growing to 1.25 m (4 ft) with grey-green leaves 5-7 cm long and 6 mm wide, broader at the top than the base. They stand away from the hanging flowers and show them off well. The flower stem curves down just above the flowerhead making it face towards the ground. The flowerhead is bowl-shaped and about 7 cm across. The bracts have a silky sheen and are suffused with russet or rose-mahogany shades. It grows wild on mountain slopes near Ceres, where snow is not uncommon in winter.

P. pityphylla PINE-LEAFED PROTEA

This little protea is similar to *P. nana*. It has needle-like leaves which are a good deal longer

◁ *Protea rupicola* has a domed head of flowers.

The Sugarbush Protea *(Protea repens)* bears flowers of sculptured beauty. ▷

than those of *P. nana*, being up to 7 cm in length. They spread back from the flower. The flower-head measures about 5 cm across and the bracts spread wider making a flatter bowl than is the case in *P. nana*. They are mahogany on the outside and shaded from ivory at the centre to russet at the tip, on the inside. The central dome of flowers is mahogany and tan. The plant is rather scrambly in manner of growth. It flowers in mid-winter.

P. pulchra GLEAMING PROTEA
(P. subpulchella)
This is one of the most attractive of the proteas of medium size. It grows to 2 m (6 ft) but can be kept to smaller size by pruning, after the flowers fade. It is bushy in habit and bears flowers rather like those of the sugarbush (*P. repens*). They can be distinguished from those of the sugarbush by the fact that the bracts of *P. pulchra* are more rounded at the tips and edged with dark fuzz. The bracts have a glistening texture and are subtly shaded, generally pink or rose, but sometimes with ice-green or lime-green suffusions. There is also a white form of this one. The leaves are long and slender, and the bush is a handsome sight from winter to early spring when it is covered with flowers. It is hardy to fairly severe frost.

P. repens SUGARBUSH, SUIKERBOS
(P. mellifera)
Twenty years ago this species used to be common in many parts of the south-western Cape and towards George but it is now comparatively rare. The name, so well known in the song "Suiker-bossie" is descriptive of the fact that it has quantities of sweet nectar which was used by early colonists as a sweetening in food.

The plant grows to 2-3 m (6-10 ft) and is erect in form. It is rather large for the small garden. The leaves are long and slender and the flowerheads are quite distinctive—rather stylised in form, with a V-shaped flower composed of pointed overlapping bracts of different shades. In some areas they are pale green and in others they are white but often they are pink to rose. Each flowerhead measures about 12 cm in length and 6-8 cm across. It flowers from late autumn to spring and tolerates a good deal of frost, and fairly alkaline soil.

P. rouppelliae DRAKENSBERG PROTEA
This protea is to be found in the summer-rainfall areas from the mountains of the eastern Cape to Natal, Swaziland, the eastern Free State and the Transvaal. It is more like a tree than a shrub and can be used as a specimen tree in a small garden or as a background plant in a large one. It grows to 3-4 m (10-13 ft). The upper stems are thickly clothed with attractive grey-green leaves about 12 cm long and fairly broad. The flowerhead is large—about 15 cm long and 10 cm across. The bracts of the flowerheads are widely spaced, silky in texture and shaded from cream at the base to pink or rose at the top. The central dome of flowers is also attractively shaded from palest lemon at the bottom to dusky rose at the top. The flowering time is summer. Like many plants which grow on mountain slopes where cold is often intense, it is damaged not by low temperatures but by dry conditions. Although in nature it is generally found growing in acid soil it shows some tolerance of alkaline soil.

P. rupicola
This protea which occurs on mountain slopes in the Tulbagh and Stellenbosch districts grows to between 1 and 2 metres in height. It has a gnarled appearance and bears unusual flowers in late spring and summer. The flowerhead measures about 7 cm across and is composed of a high-domed mass of silvery-tipped flowers with contrasting red stigmas. The surrounding bracts are short and frame the central part like a halo.

P. scolymocephala SMALL GREEN PROTEA
This protea produces dainty little flowers which are ideal for long-lasting arrangements and corsages. The bush grows to 1 m (3 ft). The leaves are long and slender, only about 6 mm wide and massed along the stems, partly obscuring the flowers. The flowerhead measures only 4 cm across and is composed of bracts of lime to ice-green arranged in the form of a bowl, with a central mass of flowers of the same colours. It flowers from late autumn to spring.

P. speciosa BROWN-BEARDED PROTEA
This beautiful protea was grown in England at the beginning of the nineteenth century and it certainly would be worth growing even if it produced only one flower per year as the flower is exceptionally lovely. The plant grows to 1.25 m (4 ft) and has broad leathery leaves. The flowerheads are about 12 cm long and never open wide. They are composed of bracts of pale to deep pink tipped with an attractive fringe of cinnamon to brown

◁ Oleander-leafed Protea *(Protea neriifolia)* produces handsome flowers of subtle colours.

Protea sulphurea is an unusual protea for gardens large or small.

silky hairs. There is a creamy-yellow one also. Its flowering time is late autumn to early spring. This species is similar to *P. grandiceps* but its leaves are not as grey, and they narrow down at the base. Also the lower bracts are bearded whilst in *P. grandiceps* only the upper bracts have a beard. It resists cold well.

P. speciosa var. **angustata**
This variety is sometimes referred to as *P. patersonii*. It is not a large plant as it seldom grows to more than 1 m (3 ft) in height and spread. Its flowers are made up of plush bracts of a charming shade of pink, each one attractively fringed with a beard, cinnamon to brown in colour. It flowers in late winter and spring.

P. stokoei STOKOE'S PROTEA
This plant which bears flowers somewhat like those of *P. speciosa* grows to 2 m (6 ft). It has leaves which are broad and oval, and the bracts are different in shape and have a shorter beard than those of *P. speciosa*. It flowers in winter. Mr T. P. Stokoe after whom it is named, certainly deserved this lovely memorial, as he was an intrepid mountaineer and plant collector until his death, fairly recently, at the age of ninety. He celebrated his 90th birthday climbing a mountain. This species should be watered well throughout the year.

P. sulphurea
Grows to 1 m (3 ft) and spreads across more than this. The leaves are small and leathery, broader at the apex than at the base, and the flowerheads are fairly small and open to form a bowl about 7 cm across. The bracts open wide in the mature flower and frame the dome of flowers, which are shaded from pale lemon-yellow at the base to cinnamon at the top. The bracts themselves are also prettily shaded from lemon to brownish-rose on the inside, and on the reverse they are pale yellow neatly edged with rose. It stands frost but needs an abundance of moisture. Its flowering time is autumn and early winter.

P. susannae SUSAN'S PROTEA
Grows to 2 m (6 ft) and has grey-green leaves about 12 cm long and fairly narrow in proportion to their length. The flowerheads which appear in winter and spring are 10 cm long and 7 cm across. The bracts are shaded from rose at the top to brown at the base. When young, the buds are very pretty indeed but as the flowers mature they tend to become rather straggly in appearance and should be cut off the plant with a length of stem to keep the bush neat and compact. A characteristic of this protea is the rather unpleasant smell the stems and leaves give off when cut. It appears to be tolerant of alkaline soil

P. venusta SWARTBERG PROTEA
This protea spreads across 2 m when mature but it is not a tall-growing species. The obovate leaves are 5-6 cm long and 2 cm broad and clothe the plant densely. The flowerheads are elegant and very pretty. The central cone of flowers is gleaming white, and the surrounding bracts are charmingly coloured from alabaster at the base to rose-pink at the top. The flowerheads measure about 6 cm in length and are cup-shaped. The flowering time is summer.

PYCNOSTACHYS URTICIFOLIA BLUE BOYS, PORCUPINE SALVIA
DISTRIBUTION: Grows in parts of the eastern and northern Transvaal and in Rhodesia.

DESCRIPTION: This is a rangy shrub growing to 2.5 m (8 ft). The leaves are broader at the base than the apex and variable in size, being larger near the base of the plant than near the top. The flowers which are blue, mauve or white, are arranged in conical spikes at the ends of stems. Those with blue flowers are the most desirable as the colour is a deep azure blue. Unfortunately the flowers do not all open at the same time but the shrub is nevertheless showy when in flower. Plant it towards the back of a shrub border so that its bare lower stems are hidden from view. It flowers in autumn and part of winter.

CULTURE: It can be grown from seeds or cuttings. Seeds sown in spring will produce flowering plants 18 months later. It is quicker, however, to raise plants from cuttings. This plant does best in gardens where frosts are not severe. In cold gardens it is apt to be cut down by frost, but the plants grow up quickly again and will flower within a few months.

RAFNIA OVATA RAFNIA

DISTRIBUTION: Its natural habitat are mountain slopes north and east of Cape Town.

DESCRIPTION: Rafnia is an upright-growing evergreen shrub to 2 m (6 ft) with leathery leaves much broader at the base than the apex. In early spring it bears pea-shaped flowers of bright yellow at the tips of the branches. *R. thunbergii* is another species worth cultivating. It is about 1 m (3 ft) high and bushy in form with soft green leaves 2-3 cm long and 3 mm wide. It bears spikes of yellow pea-shaped flowers, each spike measuring 12-20 cm in length. The individual flowers are only about 12 mm long.

CULTURE: This is a hardy shrub which stands both drought and frost when once it is established. It should be watered in winter to encourage flowering.

RHIGOZUM OBOVATUM KAROO GOLD, GEELGRANAAT

DISTRIBUTION: Occurs in dry parts of the country, the Little Karoo, the southern Orange Free State and parts of the Transvaal and Rhodesia.

DESCRIPTION: This is a shrub to bring life to gardens in dry areas where water is in short supply. When not in flower it is not worth looking at, but when it flowers in mid-spring it is a fine sight in its dry surroundings. It grows to 2 m (6 ft) and has tiny greyish leaves about 6 mm long, arranged in threes. The flowers are bright yellow with rounded crinkled petals. Plant it at the back of other shrubs so that its leggy base is hidden from view. *R. brevispinosum* is another species worth trying in dry areas.

CULTURE: In gardens which have a heavy rainfall, either in winter or summer, it is advisable to plant this shrub in well-drained soil, as it may not survive too much water about its roots. It tolerates fairly severe frost.

RHOICISSUS TOMENTOSA WILD GRAPE,
(R. capensis) MONKEY ROPE, BOBBEJAANSTOU

DISTRIBUTION: Can be found in forests from the south-western Cape east into Natal and also in the eastern Transvaal.

DESCRIPTION: This is an evergreen climbing plant which clings to any support offered. It does not have spectacular flowers but the leaves are decorative throughout the year. They are something like those of a vine in shape but are dark green and glossy on the upper surface with a paler and somewhat hairy undersurface. The new leaves are coloured claret and fawn. Under ideal conditions it produces bunches of grape-like fruits from which a tasty jam or jelly can be made.

CULTURE: Plant it in soil in which there is plenty of compost. In dry inland gardens it will do best if grown in partial shade. It should be watered regularly throughout the year, although when once established it will stand a good deal of drought and frost.

ROTHMANNIA SCENTED CUPS, SEPTEMBER BELLS

DISTRIBUTION: Grows wild in many parts of the eastern Cape, Natal, and warm regions of the Transvaal.

DESCRIPTION: These plants, known for many years as gardenias, have now been reclassified as *Rothmannia*. The two most suitable for the garden are large shrubs or small trees with handsome foliage and sweet-scented flowers.

Karoo Gold *(Rhigozum obovatum)* grows in dry places.

CULTURE: They are slow-growing plants but worth trying in districts where the climate is suitable. Plant them in holes to which plenty of compost has been added and water them well and regularly to encourage faster growth. They are not hardy to severe frost, and, in gardens which have frost, they should be planted in a protected position. They do best in warm or subtropical gardens.

R. capensis SCENTED CUPS
(Gardenia capensis)
Grows to about 3 m (10 ft) and has oval, glossy leaves. The sweetly-scented flowers are made up of a short wide tube opening up to 5 petals which curve back. They are ivory-white and the throat of the flower is spotted with maroon. The flowering time is summer. The flowers retain some of their scent when dry and are useful for making *pot-pourri*. The flowers are followed by hard round fruits the size of a small apple.

R. globosa SEPTEMBER BELLS
(Gardenia globosa)
Is a rounded shrub or small tree to 5 m (16 ft) with dark green oval leaves, and sweetly-scented, bell-shaped ivory flowers. This handsome shrub was grown in Europe more than a hundred years ago and it is worth a place in gardens in Southern Africa. It flowers in September.

SENECIO TAMOIDES CANARY CREEPER
DISTRIBUTION: Can be seen between bushes and trees from the eastern Cape into Natal and in warmer parts of the Transvaal.

DESCRIPTION: This is a quick-growing climber which twines by means of its long shoots and holds itself up on any support available be it a trellis or another plant. Its leaves are pale green, very broad at the base, and with angles somewhat like those of an ivy. It bears clusters of yellow flowers in autumn which are very striking. Each little flower is like a little yellow daisy but the clusters are large—about 12 cm across.

CULTURE: It grows extremely quickly, particularly in regions which are warm and where there is an abundance of moisture, but thrives also in dry gardens. In gardens which have frost it may be cut down to the ground but this does not matter as it keeps it from becoming too rampant a plant. It grows up again rapidly in spring and will be full of flowers again in autumn. In hot, inland gardens it seems to do best if planted where its base is shaded by shrubs or low-growing trees.

SERRURIA BLUSHING BRIDE AND OTHERS
DISTRIBUTION: Is limited in distribution to a few areas of the south-western Cape.

DESCRIPTION: Serruria species are of great decorative value in the garden and in flower arrangements. The leaves are finely divided and are soft in texture, somewhat like very slender pieces of green or grey coral in appearance. The flowers are small tubes crowned by silky hairs and enclosed in decorative bracts of different shades. They appear in winter and early spring. Serruria is a member of the protea family and like other members of this family they require special growing conditions.

CULTURE: Make holes 2 feet deep and across and fill these with acid compost and peat. Water the plants thoroughly from autumn to spring, and sparingly during the summer months. They grow best in soil which drains readily rather than in a heavy clay. In gardens which have severe frost they should be planted in a protected position.

S. adscendens
Grows to about 1 m (3 ft) in height and spread. It has small flowers of pale ivory to cyclamen, in clusters 2-3 cm across. It is not a showy plant for the garden but useful in providing sprays of flowers which last well in arrangements.

S. aemula CINDERELLA
This plant is bushy in habit growing to 1 m (3 ft) with heads of flowers carried in clusters at the ends of the stems. Each flowerhead measures 2-3 cm across and is made up of rose-coloured tubes with silvery-grey hairs at the top, closely crowded together and enclosed inside bracts of alabaster flushed with green, pink and rose. At first only the hairs at the top show through the opening at the top of the bracts, but later these decorative bracts open wide to reveal the flowers, which in turn open to make a pompon of rose and grey. They last well on the plant and in arrangements.

S. barbigera SILKY SERRURIA
Grows to a height of 60 cm (2 ft) or more and bears rounded clusters made up of heads of flowers of silvery-grey and pink to cyclamen, which look most effective above the delicately cut foliage.

Cinderella *(Serruria aemula)* bears dainty flowers.

The Silky Serruria *(Serruria barbigera)* looks charming in arrangements.

Blushing Bride *(Serruria florida)* has flowers of fragile loveliness.

The flowers of Grey Serruria *(Serruria pedunculata)* are effective in arrangements.

S. burmanii SPIDERBUSH, SPINNEKOPBOS
Is about 1 m (3 ft) in height and spread and has dainty foliage surmounted by small elegant little heads of flowers of pink to rose.

S. florida BLUSHING BRIDE, PRIDE OF FRANSCHHOEK, TROTS VAN FRANSCHHOEK
The blushing bride, with its wide spreading bracts which look as though they have been fashioned from alabaster, is undoubtedly the most beautiful of the serrurias. The flowers carried in tight heads are made up of little pink tubes crowned with silvery hairs. This is one of the prettiest of flowers for small arrangements and corsages and, in addition to being decorative, the flowers last for a long time. In fact when they are dry they are still pretty. They flower from autumn to spring when flowers are scarce, which is another point in their favour. The young plants are decorative but as they grow older they become woody at the bottom and new plants should be set out every four to five years. To look effective in the garden it is a good idea to plant two to four plants close together so that they grow into and through each other.

S. pedunculata GREY SERRURIA
(S. artemisiaefolia)
This species is densely clothed with attractive, lacy foliage of a greyish-green hue. It bears flowerheads about 4 cm across which look like silvery-grey pompons, and which are effective in the garden and in arrangements. It grows to 1 m (3 ft) in height and spread. The flowering period is spring.

STOEBE PLUMOSA SLANGBOS
DISTRIBUTION: Grows wild along the roadsides and lower mountain slopes north and east of Cape Town.

DESCRIPTION: This shrub does not produce flowers of any merit but it is a useful shrub because of the colour of its leaves. It grows to 1.25 m (4 ft) and spreads across more than this, but can be kept trimmed back to much smaller size. The stems and the minute leaves are ash-grey. The plant makes a fine contrast when planted between shrubs with green leaves. The stems are most useful in providing material for arrangements, not only because of the colour of the leaves but also because they last for weeks.

CULTURE: This plant grows readily even in poor soil and once established it will stand long periods of drought. It is also able to resist fairly severe frost. It should be trimmed once or twice a year.

SUTHERLANDIA FRUTESCENS CANCER BUSH, KANKERBOS, GANSIES KEUR
DISTRIBUTION: Can be found in many parts of the Cape Province up to South West Africa.

DESCRIPTION: It grows rapidly to 1.5 m (5 ft) and has a graceful appearance for the leaves are divided into small oval leaflets which are grey in colour. The flowers which appear in spring are very showy. Each flower is about 2-3 cm in length and made up of two sections which look like an open mouth with a recurved upper lip and a keel at the bottom. They hang down in graceful clusters and the bright tomato-red colour of the flowers shows up beautifully against the grey of the leaves. As the flowers fade, the plant becomes festooned with large, inflated, shiny, translucent seed-pods coloured green and russet-brown. Often plants are covered with the colourful flowers and attractive pods at the same time. It tends to become straggly with age and should be trimmed back a little each year after flowering. When plants become woody they should be removed and new ones planted in their place. It is a fine plant to grow towards the back of a flower border or interplanted with other shrubs in a shrub border, or lining a drive.

The common name of cancer bush is derived from the fact that early settlers used a decoction from its leaves to try and arrest the progress of cancer. The Hottentots used an infusion of the leaves to wash out wounds and internally, for fevers.

CULTURE: This quick-growing shrub grows readily from seeds or cuttings. It does well in poor soil and will endure long periods without water but should be watered, if possible, during winter and early spring. It also stands a good deal of frost.

SYNCOLOSTEMON DENSIFLORUS PINK PLUME
DISTRIBUTION: Grows wild from the eastern Cape to Natal.

DESCRIPTION: It reaches a height of 1 m (3 ft) and has an angled stem, distinctly grooved. The leaves are small, oval and aromatic. The flowers of pale pink are carried in showy clusters at the ends of stems and are most profuse in summer. Each

◁ A Serruria cultivar known as 'Maid of Honour'.

flower is tubular, ending in two lips rather like a salvia.

CULTURE: It grows quickly even in poor soil, but flowers well only if given some attention. It will stand drought but it is advisable to water it regularly from spring until flowering time. The plants may be cut down by frost, but, if they are, they usually grow up again quickly and flower in due season.

TECOMARIA CAPENSIS CAPE HONEYSUCKLE, CAPE TRUMPET FLOWER

DISTRIBUTION: Is widespread from the eastern Cape through the Transkei and Natal to the warmer parts of the Transvaal.

DESCRIPTION: This is a rewarding and decorative evergreen shrub which grows to 2 m (6 ft) and more. It has attractive shining, dark green leaves divided into leaflets with serrated edges. The flowers appear in summer and autumn and look striking as a background to the blue of plumbago. Each flower is comprised of a curved tube opening to a face with recurved petals with the stamens showing prominently between them. It is a good plant for an informal hedge or windbreak. If it grows too large it should be pruned back. The variety with lemon-yellow flowers known as *Tecomaria capensis* var. *lutea* is even more decorative than the species with orange flowers.

CULTURE: This hardy shrub is able to endure long periods of drought, but it flowers best if planted in good soil and watered fairly regularly. Severe frost may cut back the plant but it comes up again quickly and will flower by summer.

THUNBERGIA ALATA BLACK-EYED SUSAN

DISTRIBUTION: Grows wild in warm parts of Natal and the eastern Cape.

DESCRIPTION: This quick-growing climber makes a splendid show with its masses of orange flowers in summer and autumn. Each flower measures 2-3 cm across and has a dark, mahogany-black throat. The leaves are small and heart-shaped or triangular in form. The plant should be provided with a trellis on which to climb or it may be planted at the top or bottom of a bank and trained to scramble over it. Several hybrids with flowers of fawn, yellow or cream have been developed from this species It is a cheerful plant for making a quick cover anywhere in the garden and can be planted in tubs on a sunny patio too.

CULTURE: Black-eyed Susan grows very rapidly in places where summers are warm, and even when it is cut back by frost it generally grows to several feet by the time it is due to flower again in summer. When grown in the winter-rainfall area it should be watered in summer.

For a gay show in autumn plant Canary Creeper *(Senecio tamoides)*. (See page 164.)

Slangbos *(Stoebe plumosa)*. Its silver-grey leaves and stems are decorative in the garden and in arrangements.

Black-eyed Susan *(Thunbergia alata)* is a cheerful scrambling plant to cover a bank or train up a wall or trellis.

Cape Honeysuckle *(Tecomaria capensis)* makes the garden gay in summer and early autumn.

Cancer Bush *(Sutherlandia frutescens)* is a quick-growing plant even in poor, sandy soil.

IV

Part IV
Trees

◁ The Kaffirboom *(Erythrina caffra)* produces its spectacular flowers in late winter.

Trees

Trees play an important part not only in the garden but in beautifying the landscape, be it flat, undulating or mountainous. They introduce colour, height, density and form, and can transform what might be an uninteresting and monotonous scene into one of great beauty. To appreciate this fact one has only to compare some of the treeless areas of our countryside with the bushveld, and with the natural forests in parts of the Cape, Natal, the Eastern Transvaal and the eastern highlands of Rhodesia.

The beauty of the English landscape has always excited the imagination of poets, writers and painters, and enthralled visitors to that country, but few who travel through the lovely English countryside realise that much of this beauty is man-made. The park-like appearance of a great deal of England is due to the enthusiasm, energy and enterprise of the owners of large estates who, in the eighteenth and nineteenth centuries, planted not hundreds, but thousands of trees on their farms and estates. Even today trees are being planted by the million in England. School children are encouraged to take part in tree-planting programmes in their neighbourhood, and thus the beauty of the countryside of England is assured for generations to come.

With our wide, open spaces there is no practical reason why land should not be set aside so that tree-planting could be done by school children all over Southern Africa. It is during these formative years of their lives that we should engender in them a love of trees and a desire to improve the countryside. Many children would find the planting and care of trees a tremendously exciting project, particularly when done communally. There is little doubt that boys and girls who plant a tree when they first start school and who tend it for the ten or more years they are at school, will automatically know something about trees and will continue to be interested in tree-planting in later life. The words used by C. J. Rhodes, when Prime Minister of the Cape Colony seventy years ago, are as applicable now as they were then: "Whoever has the foresight to plant trees is creating a monument for himself and providing capital for his successors, while the gain to posterity is incalculable".

Whilst landowners in England for generations past were beautifying their landscape by planting trees we, in South Africa, were despoiling our natural heritage by felling them, and not replanting. It is indeed deplorable that we have for so long been oblivious to the necessity for tree-planting on a large scale, and it is even more deplorable that, in constructing our new highways,

Trees and shrubs add sparkle to the garden from season to season. *Podalyria calyptrata* with mauve flowers in the foreground, and behind is *Alberta magna* with red flowers.

many trees and beautiful old avenues have recently been destroyed.

South Africa is not richly endowed with natural forests, but three hundred years ago there were far more trees in all parts of the country than there are today.

Jan van Riebeeck, who landed in South Africa in 1652, found the Cape well-wooded. In his journal he wrote that near the fort there were fine, thick, and fairly large trees, somewhat like the beech or ash, and that the forests at Hout Bay near Cape Town were the "finest in the world". But by 1679, when Simon van der Stel became governor, many of the trees on the slopes near the fort had already been felled. Early travellers and naturalists who visited South Africa in the eighteenth and nineteenth centuries described having come across forests in many places where now there are hardly any trees at all. One of these explorers describes how he found forests near Port Elizabeth extending over thousands of acres, down to the coast, and another was enchanted by the wide bands of acacias, karee and indigenous willows which clothed the banks of the Orange River, where now few trees are to be found.

It has been estimated that more than half of the natural forests of South Africa have been destroyed since Europeans first settled in the country, and a great deal of other natural vegetation has also disappeared.

Much wood was used by early settlers for building their houses and wagons, for fencing posts and as fuel. Most of the trees originally felled for these purposes were those growing in the forests near the coast, but, after the discovery of diamonds in 1869, tens of thousands of thorn trees were cut to provide fuel and timber for the mines at Kimberley. This happened again when gold was discovered in the Transvaal; trees were felled far and wide and carted to Johannesburg for the same purpose.

Fortunately for the Western Cape, Simon van der Stel, who was governor from 1679-1699, encouraged tree-planting. He records that 16,000 oaks were planted on the slopes of Table Mountain, and many more in other areas near Cape Town. Unfortunately, his splendid example has not been followed by others. It is unfortunate, too, that little interest has been shown in regenerating forests or groves of native trees, and that gardeners tend to plant only exotic trees rather than those native to Southern Africa. This is strange for some of our flowering trees, such as the kaffirboom with its brilliant scarlet flowers, and the Cape chestnut with its trusses of orchid-like flowers, can be numbered amongst the most beautiful flowering trees of the world, whilst others, which do not produce spectacular flowers, are well worth growing because of their lovely foliage or handsome form. Even the despised "doringboom" can transform a desolate piece of countryside into a charming landscape. Where climatic conditions make it difficult to grow other kinds of trees, the "doringboom" proves a decorative and useful one, providing shade and shelter for man and beast.

In his foreword to the book *Indigenous Trees of the Cape Peninsula* by M. Whiting Spilhaus, Field-Marshal Smuts writes as follows: "It is one of the calamities of our Cape Flora that our indigenous plants have been systematically neglected in favour of imported plants. This misplaced favouritism applies especially to our trees" . . . and later, in the same foreword, he adds: "Nothing is more heartbreaking than to see our mountain sides becoming more barren every year, and one asks how long the glory of our mountain scenery will continue, and the mountains remain the main sources of our water supply. A new barbarism seems to be invading our natural scenery and destroying all that unique character and those beautiful forms of our southern flora".

It is over twenty years since Smuts wrote those words and the conditions he deplores have not changed. We still continue to grow exotic plants rather than those native to our country, and there has been no large-scale replanting of indigenous trees to take the place of those destroyed.

TREES AND THE LANDSCAPE

We have noted that the beauty of the countryside depends largely upon the numbers of trees planted. For several decades the tree-planting that has been done has been limited to the planting of exotics, mostly eucalypts and pines.

I am not against the planting of alien trees, and recognise their value, both from a point of view of enhancing the beauty of the countryside and as commercial timber. The eucalypts and pines have provided timber for the mines, for making boxes required for the packing of fruit, for building and for firewood. On the Cape Flats, after much of the indigenous vegetation had been destroyed, exotic acacias were planted to control the drifting sand. These have provided firewood for countless households, and the natural compost formed where they grow has added to the fertility of the soil. In providing fuel these exotic acacias have certainly served a purpose, for had this firewood not been readily available, the shrubby flora of the Cape Peninsula would have been completely denuded by now. These acacias have, however, proved a noxious weed, as they spread with great rapidity and have already invaded large areas and choked out the natural vegetation, which very often consisted of beautiful proteas. More use could and should have been made of native flora to prevent the drift of sand and to combat wind erosion.

We must continue to grow trees from other countries but, with them, we should grow our native trees more extensively to beautify the landscape. They should be planted

(1) in private gardens and public parks,
(2) along the streets and roads of villages and towns,
(3) on farms, and
(4) along national highways.

(1) **Trees in gardens and public parks:** In countries such as South Africa and Rhodesia, where the sun shines brightly on most days of the year, the advantage of having trees to create shade is obvious. In addition to providing shade for the house, trees should also be planted more extensively to provide shade for other plants, many of which grow better if they are not subjected to bright sunshine all day. Very often, also, the shade cast by a tree reduces the amount of watering necessary for the plants growing in its shade.

(2) **Trees along streets:** Examples of how trees have beautified towns in Southern Africa come readily to mind. There are the jacarandas in Pretoria, a diversity of trees in Johannesburg and Salisbury, flamboyants in Durban, wild figs and kaffirbooms in East London and oaks in Stellenbosch. In addition to beautifying towns, trees lining roads and streets add considerably to the comfort of the inhabitants, by creating shade and shelter. It seems strange that most of the trees planted in the streets of our towns should be exotic species. Were a survey to be made, it would probably reveal that a greater number of trees native to Southern Africa have been planted in public parks and in the streets of towns in Australia, than in towns in South Africa. There, trees such as our keurboom (virgilia), dais, and the Cape chestnut, have been grown for many years.

(3) **Trees on farms:** Most farms have some trees to provide shade and shelter around the home-

stead, but there are several other reasons why there should be more extensive tree-planting on farms. First, stock need shade from the sun in hot weather, and shelter from cold winds during winter. Groups of trees planted to shade the troughs where stock drink would help to reduce evaporation of the water, as well as provide the necessary shade for the animals; and clumps of trees in the paddocks where they graze would give essential shelter throughout the year. In addition to planting for this purpose, trees should be planted on farms to produce firewood. In many parts of Southern Africa the precious manure, which should be gathered and dug under the earth to enrich it, is burned by farm labourers because they have no other fuel. A third reason why more indigenous trees should be planted is that some of them have value as fodder. The camelthorn *(Acacia giraffae)* is one of the many native acacias which are useful in this way, and there are others, such as the karee *(Rhus lancea)*, which are palatable to stock. Trees also encourage the increase of wildlife in the form of buck and birds, many of which consume the insects which damage crops; and they provide food for bees and so increase the amount of honey available. A fifth, and most cogent reason for the planting of more trees on farms is because of their usefulness in checking erosion. The roots of trees help to bind the soil, and the leaves which fall form a covering to the top soil thereby minimizing both wind and water erosion. The prevention of wind erosion is a very real and ever-present problem in parts of our country, and there is no doubt that every year more and more top soil is being blown away, and that the desert is steadily advancing eastwards.

In areas where wind erosion is a threat to the land, it is advisable to establish windbreaks in the fields cultivated for crops and in grazing paddocks. Windbreaks are also often necessary to protect homesteads and orchards. Where trees are planted to protect a house or garden from the wind, the trees should not be planted too close to the building or garden. Where space allows, they should be planted at a distance of not less than one-and-a-half times the height they are likely to attain.

Finally, if farmers planted more trees, the whole landscape of South Africa would be changed, and it is a pity that some organisation is not formed to follow the fine example set by the Van der Stels three hundred years ago, to encourage the planting of trees in all parts of the country.

Easy-to-grow annuals decorate the ground on this long drive which is enhanced by the trees and shrubs planted.

(4) **Trees along national roads:** A limited amount of tree-planting has been done along our main highways, in isolated patches often many miles apart. In most cases exotic trees have been used. This is a pity, as visitors to our country are more interested in seeing indigenous plants than exotic ones, and it is therefore important that roadside planting should, wherever possible, consist of indigenous trees rather than those from other lands. Moreover, the planting of indigenous trees along our national roads would also help to make South Africans more familiar with them. In Rhodesia the necessity for the preservation of wide belts of indigenous trees has received recognition. The Rhodesians have shown a greater appreciation of indigenous trees and plant them more widely than we have done in South Africa.

FACTORS TO CONSIDER WHEN CHOOSING TREES

Whether trees are being planted in a private garden or a public park, as shelter for stock, for firewood on farms, or for lining streets or national highways, there are certain factors which should be considered before choosing them. The most important of these are:

(i) Climate,
(ii) Height and spread.
(iii) Foliage and flowers.
(iv) Rate of growth.

(i) **CLIMATE:** There are wide variations in climate in different parts of Southern Africa and, although the climate on the whole is temperate, there are vast areas where arid conditions make the growing of trees difficult.

The term climate includes several factors such as the amount of and seasonal distribution of rain, the relative humidity of the air, the maximum and minimum temperatures experienced, altitude, and proximity to the sea. Of these factors the most important one affecting the distribution and growth of trees is the incidence of rain.

(*a*) **Moisture:** Throughout the world luxuriant forests are found only where the rainfall is high. In Southern Africa natural forests occur principally along the slopes of mountain ranges, particularly those fairly near the coast, and at high altitudes in the eastern Transvaal, and along the south-eastern highlands of Rhodesia, where precipitation is high.

Although many of our largest and most beautiful native trees are to be found growing in areas with a high rainfall, there are some which grow naturally in regions of low rainfall, and which are adapted to stand long periods with little water. It is interesting to note that in dry areas these drought-resistant trees do not cover the country in a uniform manner, but occur in belts. There may be dense concentrations of such trees in one area, whereas in a neighbouring area, with the same rainfall and range of temperatures, few trees are to be found. Probably the reason for this uneven distribution lies in the fact that there is underground moisture in some places and not in others, and because of the nature of the soil. Some of the trees which grow naturally in the dry areas of the north-western Free State, the western part of the Transvaal, Botswana and the north-western Cape, are the camelthorn *(Acacia giraffae)*, the wild olive *(Olea africana)* and the karee *(Rhus lancea)*. These trees are also able to resist low drops in temperature in winter, and can therefore be considered hardy in every sense of the word. Our native, drought-resistant trees are more likely to flourish in dry areas than many exotic species because they have a strong root system which penetrates deep into the soil and enables them to make use of subterranean water.

(*b*) **Temperature:** The second most important climatic factor to consider when choosing trees is the range of temperatures experienced. Some trees are resistant to severe frost, whilst others will tolerate only mild frost. It must be remembered, however, that trees are remarkably adaptable and that some species can be induced to grow and flourish in a climate which is very different from that in which they occur in nature. A baobab, for example, has been grown in the Transkei, and a sausage tree *(Kigelia africana)* was grown in the Orange Free State. In both cases the climate where they were grown is quite different to that of their natural habitat.

Some trees which may succumb to frost when young, will endure fairly severe frost once they have reached the age of three to five years. This means that if the trees are protected during the winter months, while young, they will later grow to maturity without any further protection. Generally, however, it is advisable not to plant frost-tender trees in areas where frosts are severe, and to select species which are drought-resistant for dry regions.

(*c*) **Altitude:** Altitude affects the growth of trees mainly because of the climatic conditions due to altitude. If the terrain is very high the cold may be intense in winter; on the other hand, it may also have a higher rainfall than regions at lower levels. In some parts of Southern Africa the frosts which occur on plateaux at high altitude limit the variety of trees which may be grown.

(*d*) **Proximity to the Sea:** This plays an important part in the growth of trees. Few trees are tolerant of salt-laden air and sandy, salty soil, and gardeners on the seashore are therefore restricted to the growing of trees which will stand such conditions. There is quite a considerable difference in the range of temperatures and relative humidity along the coast, from South West Africa down to the western and eastern Cape and north to Natal and Mozambique. There are, however, some trees which flourish in parts of this coastal strip.

Because of the clarity of the air inland, the rate of transpiration is higher than at the coast, and trees which may need a good deal of water to produce good growth when planted inland, usually require much less water when grown in gardens near the coast. Along certain parts of the coast of South West Africa, the sea mists which prevail for long periods of the year, keep the air cool and moist and induce good growth despite the low incidence of rain.

(ii) **HEIGHT AND SPREAD OF TREES:** When trees are to be planted on farms, for shade and shelter near the house or in paddocks, the size to which they may grow is not important, whereas in towns the height and spread of trees are of the greatest importance, whether the trees are to provide shade in a private garden or public park, or whether they are to be planted in avenues along streets and roads.

In giving the height and spread of trees in the pages which follow, I have taken the growth of trees under average conditions. A tree which grows to 30 or more metres, in a forest, where the rainfall is good, may not grow to a height of more than 10 metres, when planted in the open, or when grown where there is little moisture. Indeed some forest trees may never develop to more than shrub size if planted in areas subject to severe and prolonged droughts. Similarly, a tree which grows to about 10 metres in the lowveld of the Transvaal, may grow to much more than this in parts of Rhodesia and Malawi.

Where, therefore, I have given the height of a tree as being 10 metres, and you know that in your area it grows to considerably more than this, you will realise that I have had to make some compromise, and give figures which can be considered as applicable to the height to which the tree can be expected to grow under average garden conditions:

(iii) **FOLIAGE AND FLOWERS:** When selecting trees many gardeners think only of the flowers which a tree bears, and base their selection on the beauty of the flowers. This should not, however, be the main reason for choosing a specific tree. Few trees carry their flowers for more than three weeks in the year, and it is therefore wiser to evaluate them from the point of view of foliage rather than flowers. Secondly, consider their form or shape, and let the beauty of the flowers take third place. Evergreen trees are in leaf throughout the year, and deciduous ones lose their leaves for only a short period. The foliage therefore contributes far more to the beauty of the garden, park or street, than the flowers which last for a mere three weeks each year.

There is great diversity in the shapes of trees. They may be pyramidal in form, or tall and slender; they may have a round or an oval top, or they may be spreading or drooping in shape. There is an even greater diversity in the shapes and texture of the leaves of trees. For example, some trees have leathery, glossy leaves densely arranged; others may have soft palmate leaves, which are less dense, and others may have leaves which produce a light, feathery effect.

(iv) **RATE OF GROWTH:** The rate of growth of trees is an inherited characteristic shared by all members of a species. Some are naturally slow-growers whilst others grow quickly. This applies to exotic trees as well as to indigenous ones. The belief which is widely held, that all of our native trees are slower in growth than those which are exotic, is a fallacious one. Trees such as the anaboom *(Acacia albida)* and the keurboom *(virgilia)* have been known to grow to four or five metres (12-16 ft) in three years.

What undoubtedly affects the rate of growth of trees is the kind of soil and situation in which they are planted. Hard, stony ground deficient in humus will slow down the rate of growth of any tree, and so will a scarcity of moisture. To promote good growth one should plant trees in good soil and one should water them as abundantly as possible. This does not mean sprinkling the surface of the soil; it means giving them 45 litres (10 gallons) or more a week during dry weather if they are in active growth. I have found that the best way to water trees during dry periods is to measure the water, either by filling buckets and tipping the water out into shallow "basins" made around the young trees, or by timing the rate of flow of the water from the hose, and then estimating how long the hose should remain in one place to provide the tree with 14-18 litres (3-4 gallons) at a time. In many parts of the country where a shortage of water makes it impossible to water young trees liberally, the ground around the trees should be heavily mulched with straw, or even stones, to prevent evaporation of water from the soil. More young trees die through lack of water than for any other reason. After three or four years, when they are established, many of them will survive long periods without water, whilst others will slow down in growth if they are allowed to become dry for a long period. For this reason, gardeners living in an area of low rainfall, where there is a scarcity of water for irrigation, should choose only those trees which are drought-resistant. Where, however, there is an abundance of water for irrigation a wide range of trees can be encouraged to grow.

The nature of the soil is most important, also, in promoting good and fast growth, particularly of young trees. In gardens where the soil is poor it is advisable to make holes a metre deep and across, and to fill these up with compost, or compost and manure, added to some of the soil previously removed from the hole.

If, perchance, the trees you have chosen for your garden all happen to be naturally slow-growing, fill up the spaces between them with quick-growing shrubs and climbers which can be removed later when the trees have grown.

PROPAGATION OF TREES

Generally, it is wiser to purchase trees from a nursery as in this way one saves two or more years of waiting for the tree to reach the size it has attained in the nursery. If, however, you wish to propagate your own trees, sow the seeds in spring or summer in soil which is rich in humus, and see that they are watered to encourage germination. Many of the seeds with hard shells may take weeks or even months to germinate. Germination of these can be speeded up by filing a notch in the hard outer coat of the seed, or by soaking them in water for some days. Some trees can be propagated easily also from cuttings or suckers, whilst others are difficult to root from cuttings.

PLANTING

When purchasing trees from a nursery do not expect to procure trees of large size in containers, as the container restricts the spread of the roots and this, in turn, limits the top growth. In the nursery deciduous trees are often grown in rows in the ground and not in containers. These are lifted and despatched only in winter when they are dormant. In this case large specimens can be transplanted from nursery to garden.

Trees raised in containers may be transplanted at any season of the year, but many gardeners prefer planting them at the beginning of the rainy season rather than during the dry season. When moving trees from a tin, first wet the soil in the tin, and then tap the tin against a firm surface all around, to loosen the soil so that it comes out in one piece about the roots of the tree. The less disturbance to the roots of trees which are in leaf at planting time, the better the tree is likely to take to its new environment. Water the plant thoroughly after planting, and keep it well watered until the roots have become established, which may take six or more weeks.

ACACIA THORN TREES AND OTHER COMMON NAMES

DISTRIBUTION: Different species are to be found growing in various parts of Southern Africa, particularly in inland districts and on the highveld and lowveld.

DESCRIPTION: Some of our acacias can be considered ornamental trees for gardens, small and large, and much more use could be made of them in gardens. They could and should be planted more extensively also on farms to provide shade and shelter for stock, and to relieve the monotony of the bare landscape characteristic of much of Southern Africa. Some of them have the additional merit of producing leaves and seed-pods which provide fodder for stock.

Acacias are easily recognisable by their feathery leaves and their balls or spikes of fluffy flowers, which are often yellow but which may be cream or white. They also all have thorns, either straight ones or short, curved, hooked ones. Some have both. Thomas Baines, one of our early explorers, grouped these trees in an amusing way. He recorded that one group were for tearing clothes, another for tearing flesh, and a third for tearing flesh and clothes both together. It is interesting and helps in identification to know that the acacias with round balls of flowers have straight thorns, or straight and hooked ones together on the same tree, whereas those with spikes of flowers have hooked thorns. There is only one exception—the Anatree *(A. albida)* which has spikes of flowers and straight thorns.

CULTURE: Acacias grow easily and many of them grow fairly quickly too. The seed should be soaked for two or three days, or else put into boiling water and allowed to remain in the hot water for several hours. This speeds up germination. They tolerate poor, sandy soils but undoubtedly grow quicker when planted in ground improved by the addition of compost, and when watered regularly during the first two years. Some species which are tolerant of dry conditions are excellent trees to use in the garden in regions where water is limited, and where temperatures are very high in summer and drop to below freezing at night in winter.

A. albida ANATREE, ANABOOM, WHITE THORN, APIESDORING

Occurs in the Transvaal, South West Africa, Rhodesia and Botswana. In its natural habitat one occasionally finds specimens of 18 m (60 ft) in height, but generally its height and spread are much less than this—approximately 9 m (30 ft). It is a quick-growing, deciduous tree, too large for the average town garden but a good background one for farm gardens and to provide shelter and shade for stock. It is also a tree worth planting at the sides of roads to give shade to travellers and to line broad avenues. The large seed-pods which curve almost into a circle provide fodder for animals. This tree has grey bark and the usual feathery acacia foliage, straight thorns in pairs about 12-25 mm long, and spikes of flowers of ivory to creamy-yellow, in late winter. All the other acacias with spikes of flowers have hooked thorns.

A. caffra KAFFIR THORN, KAFFIR-WAG-'N-BIETJIE, KATDORING

Is common in the eastern Cape, the Transkei, Natal, parts of the Transvaal, Rhodesia and Botswana. Under good conditions it grows to 8 m (25 ft) in height and less in spread, and often has a crooked, rough, black bole, but it is nevertheless graceful in appearance. The new leaves in spring are particularly pretty, and the foliage is useful for supplementing the food of stock. It has small, hooked thorns in pairs and the feathery leaves are made up of small leaflets. The creamy-yellow flowers carried in spikes in late spring are followed by long, narrow seed-pods. Once established it will stand fairly severe frost. This species is tolerant of a good deal of drought and is therefore a useful one for hot, dry gardens. Under good conditions it is quick-growing, and it should be more widely planted in gardens and as avenues along roads.

A. galpinii MONKEY THORN, APIESDORING

This is a robust species found mainly near streams in the Transvaal bushveld and further north, in Rhodesia. Under optimum conditions this deciduous tree reaches a height of 18 m (55-60 ft), and has a wide rounded canopy of feathery foliage. Generally, however, it seldom grows to more than about 12 m. It is too large a tree for the average garden in town but an excellent one for farm gardens and for providing shelter and shade for stock. It also makes a good avenue tree and should be used more extensively along our national roads. It resists a good deal of drought and moderate frost and is fairly quick-growing. The leaflets are small. The flowers, carried in spikes in summer, are creamy-yellow with a honey-like scent. It has small brown, hooked thorns which occur on the trunk as well as on branches and twigs.

A. giraffae CAMEL THORN, KAMEELDORING

Is widespread in sandy areas of the north-western part of South Africa, in South West Africa, Rhodesia, the Kalahari, and the western side of the Orange Free State and Transvaal. This is a hardy deciduous tree which in dry areas seldom grows to more than 8 m (25 ft) but where the rainfall is good it reaches a height of 12 m (40 ft). It has a graceful spreading umbrella crown, and, under optimum conditions, will cover 15 m. The bark is rough and dark in colour and the flowers which appear in late winter and early spring, are round yellow balls. The strong thorns carried in pairs are 3-5 cm long and the seed-pods are 6-10 cm long and fairly broad with a characteristic velvety covering. They are relished by stock and game, but at certain times of the year may be toxic because of the prussic acid present in them. The common name of this tree is derived from the fact that both the foliage and the flowers are eaten by giraffe—once known as the camelopard. This is a slow-growing species but worth cultivating, particularly in gardens where the soil is poor, the water supply meagre and frosts are fairly severe. It does well in alkaline soil. The camelthorn makes a pretty shade tree. On farms it should be more extensively cultivated to provide fodder and firewood, as well as shade.

A. haematoxylon VAALKAMEELDORING, BASTERKAMEEL, VOLSTRUISKAMEEL, KABOOM

Occurs in dry sandy areas of the northern Cape and the Kalahari. This little acacia seldom grows to more than 6 m (20 ft) in height and in very dry areas it often is much smaller. The leaflets are small and silvery-green in colour. The thorns are long and straight and the flowerheads are fluffy round balls. This is a tree for difficult conditions, where water is at a premium, the heat intense and the air dry throughout the year.

A. karroo SWEET THORN, MIMOSA, SOETDORING

This is the most widespread of the trees described, being found in many parts of South Africa, South West Africa, Botswana and Rhodesia. In some areas of the eastern Cape it is apt to prove a nuisance because it takes over land which could be used for other purposes. Generally it grows to about 4-6 m (12-20 ft) and has a rounded crown.

The Sweet-thorn or Mimosa *(Acacia karroo)* becomes a decorative shade tree when mature.

Where conditions suit it this species may grow a good deal larger than this. It makes a good shade tree of medium size. The light and feathery foliage is composed of very small leaflets, and the fluffy balls of yellow flowers which adorn it in summer are sweetly scented. The thorns near the base of the tree are long and strong whilst near the top they are few and small. They grow in pairs and vary in length from 1-10 cm. The name "Sweet-thorn" was given because it is thought that where this tree occurs the grazing is sweet, because the foliage and young pods are eaten by stock, or because of the flavour of the gum it exudes. During the years of colonization of our country this proved one of the most useful of trees. The wood was, and still is, widely used as fuel. It was also often used for furniture, for making wagon wheels and yokes and for fencing posts. The thick bark was made use of also by the early colonists for tanning leather, and the trees were cut down and piled close together to make kraals for their cattle. In some parts of the country they are still used for this purpose. This species is used medicinally by Bantu tribes for many purposes. It is a pleasing tree to plant for shade in the garden and is recommended for planting on farms to provide shade, shelter and fodder for stock, and for firewood. It is a fairly quick-growing tree and, once established, will stand both drought and cold. It does well in alkaline soil.

A. nigrescens KNOB THORN, KNOPPIESDORING
Grows naturally in the lowveld of the Transvaal, Swaziland and Natal, and north into Rhodesia and Malawi. It is common in the Kruger Park. This is a large deciduous tree growing to a height of 10 m (30-35 ft) or more, with a rounded top, but usually the specimens seen are smaller than this. Its leaflets are more rounded and larger than are generally found on acacias, and the thorns are carried on top of little knobs on the trunk and branches, which accounts for the common name. These thorns are hooked and strong. The flowers which appear in profusion in late winter and early spring are in spikes, brownish-red in bud and creamy-yellow when mature. It likes warm growing conditions and stands only mild frost.

A. robusta ENKELDORING, OUDORING
Can be found in Rhodesia, the Transvaal, Natal, Botswana and the northern Cape. This is an attractive thorn tree growing to 10 m (30-35 ft) in height. It has a gnarled rough bole and branches, and dainty feathery leaves. The thorns which grow in pairs are 1-7 cm long. The balls of ivory flowers which appear in early spring, often before the leaves, give off a pleasing scent. Once established it will stand both cold and drought.

A. sieberiana var. **woodii** PAPERBARK THORN, NATAL CAMEL THORN
Can be found in Natal and north into the Transvaal lowveld and bushveld and in Rhodesia and Botswana. It grows to 8 m (26 ft) and more, and has a wide, spreading flat crown of distinctive form. The bark, which is corky, flakes off in papery strips which accounts for its common name. The leaves are divided into very small leaflets and the thorns are straight and short. The flowers are white or creamy-yellow balls and not carried in great profusion. The tree is regarded as a good fodder tree, for both foliage and seed-pods are relished by animals. As with other acacias, at certain times of the year the seeds seem to contain a certain amount of prussic acid which is toxic to animals. This is a good tree for dry gardens where frosts are not severe. It is fast-growing and tolerates drought.

A. tortilis subsp. **heteracantha** UMBRELLA THORN, HAAK-EN-STEEK
This acacia is to be found in the northern Cape, the western Orange Free State, the bushveld and lowveld of the Transvaal, Botswana, South West Africa and further north into Rhodesia. Although it may reach a height of 12 m (40 ft) it seldom

grows to more than 3-6 m (10-20 ft) in its natural surroundings. The tree varies in shape considerably sometimes having a flat spreading crown, and sometimes a rounded top with branches coming up from near ground level. The foliage is very fine and grey-green and the sweetly scented flowers appear in creamy coloured balls in summer. The bark is rough and dark and the thorns arranged in pairs are of two kinds—small hooked brown ones and long white straight ones, which accounts for the common name of "Haak-en-steek", meaning hook and prick. This accounts also for the subspecies name which is Greek for "different thorns". The seed-pods which are twisted into a spiral, provide nutritious fodder for stock. A well-grown specimen of this tree is very decorative and it should be used more in gardens where growing conditions are difficult for it stands both drought and frost.

A. xanthophloea FEVERTREE, SULPHUR BARK

Is found in warm regions of northern Zululand, the lowveld of the Transvaal, Swaziland and Rhodesia, and in Mozambique. The name is derived from the Greek *xanthos*, yellow, and *phloeos*, bark. This tree is easily identified by the unusual greenish-yellow colour of the bole and branches. It is a graceful, deciduous tree which reaches a height of 12 m (40 ft). The leaflets are small and the thorns carried in pairs are straight and white. Sweetly-scented, round, yellow flowers appear in summer. The tree is worth growing because of its form and the colouring of the bark. The common name of "Fevertree" is due to the fact that it grows naturally in areas where malaria proved a scourge and was thought to have the power of inflicting this illness on those who passed near it. It will always be remembered by those who have read Rudyard Kipling's story of the adventures of the Elephant Child who travelled to the "great, grey-green, greasy Limpopo River, all set about with Fever Trees".

This tree will stand a certain amount of frost but should not be tried in areas subject to long periods of drought and severe cold. Under suitable conditions it is quick-growing.

ADANSONIA DIGITATA BAOBAB, CREAM-OF-TARTAR TREE, LEMONADE TREE

DISTRIBUTION: This unusual tree occurs from the northern Transvaal through Rhodesia, as far as Ethiopia.

DESCRIPTION: The baobab with its huge girth is not a tree for the town garden but it certainly could be used with effect in large gardens and parks. The tree has an odd appearance with a massive bloated trunk and branches which, when bare of leaves, look like roots projecting into the air. It is a deciduous tree which produces its leaves after rains have fallen in spring or early summer. The leaves of mature trees are divided into 5-7 large dark-green leaflets which spread out like the fingers from the hand. The huge flowers which appear in late spring are about 15 cm (6 in) wide and waxy-white in colour. They hang gracefully from the tree and are followed by large oval fruits with a velvety exterior measuring up to 15 cm in length and 8 cm across. These enclose the seeds embedded in pulp with an acid flavour. The pulp later becomes powdery and, mixed with water, it makes a pleasant acid drink which accounts for the common names of cream-of-tartar tree or lemonade tree. A large tree may be 12 m (40 ft) or more in height with a bole having a circumference of about this measurement, 1 m from the ground. It is surprising that this dramatic-looking tree has not been more widely grown where conditions suit it. Although they grow in dry areas these trees contain a large proportion of water.

Many are the uses made of the tree by indigenous African people. They hollow out the trunk and use it for storing their corn, and, in times past, some made homes in its branches to be safe from the prowling lions. The leaves and fruit pulp are used in parts of Africa for the treatment of fever, and some tribes cook the leaves as a vegetable. Before material for clothes became available they used fibre obtained from its bark for making clothes and nets for catching fish. Early colonists also made use of these trees. It is recorded that the hollowed-out trunk of one was used by a farmer as a dairy; another was used as a bar, by prospectors looking for gold in the vicinity where it grew; and in another, in the Caprivi Strip, a lavatory with flush sanitation was installed!

CULTURE: Although few records are available of baobabs grown from seed, it is known that given good soil, moisture and a climate that is not harsh, they grow fairly quickly. In nature they grow in soil which is not particularly good, where the rainfall is low (from 250-500 mm a year), and where frosts are never more than mild. It is recorded by Dr. L. E. W. Codd that a seedling planted by Dr. I. B. Pole Evans grew eight metres in fifteen years.

Large specimens are also to be found in Ceylon presumably grown from seed taken across by Arab traders many years ago.

AFZELIA CUANZENSIS — RHODESIAN MAHOGANY, MAHOGANY BEAN

DISTRIBUTION: Occurs in parts of the Transvaal lowveld and is plentiful in parts of Rhodesia and Angola.

DESCRIPTION: This is a handsome, deciduous tree which, under ideal conditions, will grow to 18 m (60 ft) in height but the usual height and spread is half this. It has a spreading crown and graceful form. The leaves are divided into large, shiny oval leaflets which are dark-green and which hang down prettily from the tree. The small rose-pink flowers are sweetly scented and carried in clusters at the ends of the stems. They are not showy but are followed by large seed-pods 25 cm long and 5 cm broad, which contain attractive shiny, black beans trimmed with red. These are used by some African people to make necklaces. This is a well-shaped and decorative tree for large gardens and parks and for avenue planting.

CULTURE: It grows in regions where winters are never more than mild and, although when once established it stands quite considerable drought, it grows more quickly where it receives a good deal of moisture. It is slow-growing but long-lived.

ALBERTA MAGNA — ALBERTA

DISTRIBUTION: Can be found in warm parts of the eastern Cape, Natal and Zululand, often on the fringe of forests in regions where the rainfall is good.

DESCRIPTION: This is a very decorative small tree or large shrub which, under optimum conditions in its native haunts, will reach a height of 8 m (26 ft). Generally in the garden it seldom grows to more than 3-5 m (10-16 ft). It is decorative throughout the year because of its glossy leaves and its long flowering period. The leaves are oval and pointed, up to 12 cm in length, dark green on the upper surface and paler on the underside, with a prominent central vein. The flowers of coral-red to tomato-red show up handsomely against the dark colour of the leaves and they remain on the tree for a long time. They are carried in rather flat clusters towards the ends of stems. After the flowers fade the sepals enlarge and become terra-cotta to coral-red, and very often the plant is brightened by these colourful sepals and flowers at the same time. The main flowering time is late winter and spring.

CULTURE: Alberta does best in regions where the rainfall is good or where it can be watered regularly. It is a fine shrub for coastal gardens but is not recommended for gardens which are dry for

Alberta *(Alberta magna)* showing detail of flowers and sepals.

Alberta *(Alberta magna)* showing arrangement of flowers on the tree.

long periods or where severe frosts are experienced. Plant it in a hole to which an abundance of compost has been added and water it regularly throughout the year. It is slow-growing but well worth a place in the garden where conditions are suitable. It is difficult to grow from cuttings or from seed but some nurseries now have plants for sale.

ALBIZIA FLAT-CROWN, PLATKROON

DISTRIBUTION: The most decorative of our native species grow in the warmer parts of Natal and the lowveld of the Transvaal, Rhodesia and further north.

DESCRIPTION: They are small trees which resemble the acacias. They have attractive foliage made up of many small leaflets, and fluffy flowers with protruding stamens, which enhance the appearance of the flowers. Although similar in appearance to some acacias, they are easily differentiated as they have no thorns. Our native species are not as attractive as *A. distachya,* which grows wild in Australia, and *A. julibrissin* which is native to Asia, but the species mentioned are worth growing for their ornamental value.

CULTURE: The South African albizias do best in regions where winters are fairly mild, but they tolerate long periods of drought and poor soil, and once established will stand moderate frost. They are fairly quick-growing particularly if watered regularly during their first few years.

A. adianthifolia FLAT-CROWN, PLATKROON,
(A. gummifera) MUNJERENJE

This species can be found in parts of the Transkei, the warmer areas of Natal and the Transvaal, and in Rhodesia. The flat, spreading crown of this tree makes it a decorative shade tree for the garden. It grows fairly quickly to a height of 8 m (26 ft) and, in coastal gardens and parts of Rhodesia, may grow larger than this. It is a deciduous tree and the new leaves, when they emerge in spring, are a pleasing shade of lime-green. They are divided into tiny leaflets only 12 mm long, with midribs running diagonally across. The flowers of greenish-yellow show up in October and November. This tree is suitable for small or large gardens and also for street and avenue planting. It stands moderate frost.

A. harveyi

Grows in the lowveld, Botswana and Rhodesia. It seldom reaches a height of more than 6 m (20 ft) and has pretty feathery foliage and white heads of fluffy flowers in spring. It endures long periods of drought and a good deal of frost.

APODYTES DIMIDIATA WHITE PEAR, WITPEER

DISTRIBUTION: Can be found in forests of the south-western Cape, in Natal and the Transvaal.

DESCRIPTION: Although this tree may grow to 18 m (60 ft) under good conditions in a forest, it is unlikely to reach much more than half this when grown in the garden. It is an evergreen tree with small dark green leaves which are broadly oval in form and rounded or blunt at the apex. They are somewhat leathery and glossy and make a dense crown of rounded form. This is a handsome tree with a white trunk which is slightly fluted.

CULTURE: The white pear does best in gardens where the rainfall is good and where frosts are not severe. It is not suited to dry hot gardens as under such conditions it is unlikely to grow to more than shrub size.

BAIKIAEA PLURIJUGA RHODESIAN TEAK, RHODESIAN CHESTNUT, ZAMBEZI REDWOOD

DISTRIBUTION: Can be found in parts of Rhodesia and Botswana.

DESCRIPTION: Under good conditions this tree grows to a height of 12 m (40 ft). It has a rounded top of densely arranged compound leaves. The leaflets are oval with a blunt apex and about 3-6 cm long. In late summer and early autumn it bears spikes of attractive flowers of mauvy-pink. The form of the inflorescence somewhat resembles that of the horse chestnut hence the common name of Rhodesian chestnut. Each flower has five crisped petals and prominent anthers. They are followed by large, velvety, brown seed-pods. The wood of this tree is one of the best for floors and furniture.

CULTURE: This is a slow-growing tree which occurs naturally in regions where the soil is sandy, the rainfall low, and winters are mild. It is not suitable for gardens where frosts are severe.

BOLUSANTHUS SPECIOSUS TREE WISTARIA, VAN WYKSHOUT

DISTRIBUTION: Occurs in the northern and eastern Transvaal, warm parts of northern Natal, Rhodesia, Botswana and Malawi.

DESCRIPTION: This is one of our most attractive

trees, growing to 6 m (20 ft) and sometimes much more. It has dark, brownish-grey, corrugated bark and is of graceful form, with a slender head and drooping stems of shiny leaves divided into pointed oval leaflets about 5 cm long. These fall off in late winter and often the tree is still bare of leaves when the flowers appear in September and October. The pea-shaped flowers are carried in handsome, pendant sprays about 15 cm long, and are of a lovely shade of bluish-mauve, rather like those of wistaria. When in full flower it is an outstanding sight. There is a form with white flowers, too. This is a fine tree for gardens and for street and avenue planting.

CULTURE: Once established the tree wistaria will stand a good deal of frost and drought, but it should be watered regularly and well for the first three to four years to promote sturdy growth. It is fairly quick-growing if planted in good soil and watered regularly.

BRABEJUM STELLATIFOLIUM WILD ALMOND, WILDE-AMANDEL

DISTRIBUTION: Grows naturally in the south-western Cape.

DESCRIPTION: This small tree is the first indigenous tree cultivated in South Africa. It was the one used by Van Riebeeck to make a boundary or hedge to the first settlement at the Cape. Writing in his journal in 1661 he describes the hedge, which is "growing well and will soon be high and thick". The remains of this hedge can still be seen on the ridge above the Botanic Gardens at Kirstenbosch, Cape Town.

The wild almond is a round-headed tree growing to about 5 m (16 ft) with grey-brown bark and lance-shaped leaves with toothed edges, arranged in whorls up the stem. In summer the white, sweetly-scented flowers appear in spikes about 5 cm (2 in) long, and in autumn it bears fruits rather like those of the cultivated almond in shape.

CULTURE: Although it does not appear to have been tried in cold, inland gardens it is likely that the wild almond will stand moderate frost, and, once established, it will also endure long periods without water. It makes a dense hedge or windbreak.

BRACHYLAENA DISCOLOR VAALBOS

DISTRIBUTION: Forms part of the coastal bush from the eastern Cape into Natal.

DESCRIPTION: This evergreen tree is variable in habit—sometimes growing as a shrub and sometimes as a tree with a distinct trunk. It seldom reaches a height of more than 6 m (20 ft). It is very useful for gardens near the seashore where wind and salt spray make it difficult to grow many other trees or shrubs. The leaves are grey on the underside giving the plant a pleasing silvery appearance when the wind blows. It stands trimming well and can be grown as a hedge or planted close together and allowed to grow taller to form a windbreak. It should prove a useful plant to check the drift of sand along the coast.

CULTURE: It will grow inland but it does not tolerate severe frost. It is recommended mainly for coastal gardens in windy areas.

BRACHYSTEGIA SPICIFORMIS MSASA

DISTRIBUTION: Grows naturally in Rhodesia, Malawi and further north.

DESCRIPTION: The growth of this tree is variable. It may remain stunted or it may develop into a tall tree. It is likely to grow to 9 m (30 ft) under good garden conditions. This is a deciduous tree with compound leaves made up of oval leaflets 5-10 cm long. The flowers which appear in spring are inconspicuous, but sweetly scented. They are followed by large seed-pods. The tree is worth growing because of its colourful spring foliage. The new leaves, which appear in September, are a luminous, coppery-red, pink or light tan, and a mass of trees is a very beautiful sight. This tree should be more widely planted in private gardens, in avenues and as a street tree. Closely planted it will make a decorative windbreak.

CULTURE: Although this tree grows in abundance in parts of Rhodesia it does not seem to grow readily under cultivation. Once it is mature it stands quite severe frost and drought.

BUDDLEIA SALVIIFOLIA SAGE WOOD, SALIEHOUT

DISTRIBUTION: Occurs naturally in many parts of the country, principally in Natal and the Transvaal but also in the eastern Orange Free State and Lesotho.

DESCRIPTION: This tree is decorative in late winter and early spring when it becomes smothered with large trusses of small flowers which are lilac in colour with yellow to orange markings in the centre. Apart from being attractive in appearance

In late winter Sage Wood *(Buddleia salviifolia)* produces masses of honey-scented flowers.

The Cape Chestnut *(Calodendrum capense)* has good foliage and lovely flowers.

the flowers give off a delicious scent of honey which attracts bees from far and wide. It is the first plant to scent the air in late winter. The leaves are slender, oval and pointed, rough in texture, dark green on the upper surface and grey on the underside. The plant grows to 6 m (20 ft) or more but can be kept down to shrub size by pruning each year, after it has flowered. Other native buddleias worth growing are smaller in growth and have flowers of creamy-yellow or pale orange-yellow.

The sage wood is not a striking tree but it is quick-growing and should be planted to fill up space whilst slow-growing trees are maturing. It is worth growing also for the wonderful scent of its flowers.

CULTURE: It grows well under difficult conditions, enduring both sharp frost and long periods of drought. Where frost cuts it down it usually grows up again very quickly. As it tends to become straggly with age it is advisable to trim it back every two or three years to make it more shapely.

BURKEA AFRICANA — WILD SERINGA, RHODESIAN ASH

DISTRIBUTION: Is found in the lowveld and bushveld of the Transvaal, Rhodesia and further north to Ethiopia.

DESCRIPTION: This tree is sometimes difficult to identify as its growth varies considerably. In some regions it may be no more than a straggly shrub to 3 m (10 ft) whilst in other regions it may grow as an attractive tree to 10 m (30-35 ft) in height with a grey bole. Generally it is about 6 m (20 ft) tall. The compound leaves are divided into oval leaflets each about 2-3 cm long and 12 mm broad. The flowers, which hang down on slender string-like stems like tassels, are ivory and fragrant. They are followed by large, woody, brown seed-pods. The tree is pretty in autumn when the foliage turns shades of copper and russet, and in spring, when the new leaves appear. It is a graceful tree with a somewhat flattened crown.

CULTURE: Although it is slow-growing this tree is worth trying in districts where winters are mild. It seems to do better in sandy soil than in clay, and grows larger where the rainfall is good than in dry areas.

CALODENDRUM CAPENSE CAPE CHESTNUT, WILD CHESTNUT

DISTRIBUTION: Grows in forests from the south-western Cape through the Transkei to Natal and parts of the Transvaal, and further north into Rhodesia.

DESCRIPTION: Under forest conditions the tree reaches a height of about 18 m (65 ft) but when grown in gardens it seldom grows to more than 10 m (30-35 ft). It is deciduous or evergreen depending on climate. The crown is rounded to spreading in shape and the leaves are oval, and pointed, about 10 to 15 cm long and 7 cm wide. They are of a lovely mid-green shade and slightly glossy. The midrib is clearly defined and so are the veins on the underside of the leaf, which is paler than the top surface. It is an attractive tree at any time but really striking in late spring and early summer when it is crowned with large open clusters of dainty pinky-mauve flowers. Each flower has five slender petals marked with maroon to crimson. This is a fine tree for the garden or for street or avenue planting, particularly near the coast. The word *calodendron* means "beautiful tree", and it well deserves this name. This handsome tree should be more widely planted in all regions where frosts are not severe.

CULTURE: This is a slow-growing tree, which takes six to eight years or more before it flowers. Plant it in good soil in which there is plenty of compost, and water it abundantly during its first three or four years. Once established it will stand moderate frost and fairly long periods without water. It has been grown in the Karoo and in Johannesburg and it has been planted in Australia. It does best in a mild climate with a good rainfall.

CASSIA ABBREVIATA subsp. **BEARANA** LONG-TAIL CASSIA

DISTRIBUTION: Its natural habitat is the bushveld and lowveld of the Transvaal, Rhodesia, Botswana and further north into tropical Africa.

DESCRIPTION: This is a decorative tree for the small garden or patio, or for roadside planting. It seldom grows to more than 6 m (20 ft) and has a graceful crown of drooping, feathery foliage. The leaves are made up of four to eleven pairs of oval leaflets, 3-5 cm long and 1 cm broad. The five-petalled yellow flowers are carried in clusters at the ends of stems. They are sweetly scented and appear with, or just before, the new leaves in spring. The curving stamens add to the attractiveness of the flowers. The flowers are followed by long, slender seed-pods.

CULTURE: This is a fast-growing little tree which will stand considerable drought, but it is tender to sharp frost. Once established, and if protected when young, it will, however, later endure occasional severe frosts.

CASSINE CROCEA SAFFRON, SAFFRAAN

DISTRIBUTION: This tree is to be found in the coastal forests of the Cape and eastwards into Natal.

DESCRIPTION: The saffron tree is a tall spreading one, usually evergreen, with oval leaves about 5 cm in length. They are dark green, leathery and toothed—and have a resemblance to those of the English holly. Insignificant bunches of greenish-white flowers appear in spring. These are followed by the fruits which are oval berries. The yellow pigment under the bark shines through and gives rise to the common name. The species *C. peragua (C. capensis)* is similar but smaller in size.

CULTURE: This tree is fairly quick-growing in areas of good rainfall, attaining a height of about 6 m in ten years. Once established it will stand moderate frost and drought, but it undoubtedly does better in areas where the rainfall is not less than 500 mm a year. Plant it in good soil and water it well to encourage growth. *C. peragua* (Bastard Saffron) is more tolerant of dry conditions.

CELTIS AFRICANA *(C. kraussiana)* WHITE STINKWOOD, CAMDEBOO STINKWOOD, WITSTINKHOUT

DISTRIBUTION: It grows from the coast in the eastern Cape up through the highveld to beyond the borders of South Africa as far north as Ethiopia.

DESCRIPTION: The growth of the tree shows considerable variation according to conditions. Where the soil is poor or where there is an abundance of rock and little water, it may be only 2-3 m (6-10 ft) high whilst in moist forests it will reach a height of 20 m (65 ft) and more. Under average garden conditions its height and spread will be about 10-12 m. The leaves vary too in size from 2 cm to 10 cm in length. The tree can be recognized by its leaves which are broad at the base tapering to the apex, with finely serrated edges and three distinct main veins coming up from the base, and by its smooth white or pale grey bole and branches. It is a deciduous tree with new spring leaves of delicate green turning later to dark green. The small green flowers are inconspicuous but the tree is worth growing because of its fine form. It bears small round, brown berries after the flowers fade. This is a splendid shade tree for a large garden and well suited too for avenue, street and park planting. It is unsuitable for small gardens because of its spread. The white stinkwood is not related to the black stinkwood.

CULTURE: Under normal garden conditions the tree will grow fairly quickly. Once established it stands fairly long periods of drought and sharp, but not severe, frost.

COMBRETUM ERYTHROPHYLLUM
BUSH-WILLOW, VADERLANDSWILG

DISTRIBUTION: Commonly seen near streams in Natal, the bushveld and the highveld. The species name is Greek meaning "red leaves" and refers to the autumn colours of the foliage.

DESCRIPTION: This is a small deciduous tree with a grey bole and glossy, lance-shaped leaves which turn pretty shades of russet and gold in autumn and winter. The flowers are insignificant but the tree is worth growing because of its autumn colours. The winged fruits which follow the flowers are quite decorative and remain on the tree for a long time. It is very variable in size according to growing conditions. Under normal garden conditions it will grow quickly to a height of 9 m (30 ft), and make a useful shade tree. It also makes a good windbreak or screen for a large garden or farm. Two other species worth trying in gardens where winters are not extreme are *C. kraussii* which is found in forest areas of the Transkei and Natal, and *C. zeyheri*, which is common in the bushveld of the Transvaal and in Rhodesia. They grow to a height of 6 m (20 ft) and they both have leaves which turn autumn colours. The latter has particularly large four-winged fruits measuring up to 7 cm in length. They are both fairly quick-growing.

CULTURE: It is prepared to grow under difficult conditions but remains small if planted in regions which experience long periods of drought. If planted in good soil and watered regularly until well established it grows quickly. It is tender to sharp frost when small but hardy when mature.

CUNONIA CAPENSIS
RED ALDER, ROOI-ELS

DISTRIBUTION: Grows in the Cape, Natal, the Transvaal and further north, usually in forests where moisture is present.

DESCRIPTION: This is an evergreen tree growing to 18 m (60 ft) under optimum forest conditions, but generally it reaches a little more than half of this when grown in the open. It has a crown of attractive glossy leaves composed of long leaflets with sharply toothed margins, darker on the upper surface than the under. The leaves are arranged opposite each other and have large spoon-shaped stipules. The cream flowers are carried in neat cylindrical spikes in late summer and autumn. These are followed by little brown capsules containing the seed. This is a good garden or avenue tree under suitable climatic conditions. In Europe it has been grown as a pot or tub plant for some years.

CULTURE: Where conditions are harsh the red alder remains small and makes an attractive shrub. Under normal garden conditions, if it is planted in good soil and watered regularly, it will rapidly grow into a decorative tree. It is not hardy to severe frost but may survive in a sheltered place, in cold regions.

Red alder or Rooi-els *(Cunonia capensis)* is worth growing for its pleasing foliage.

CURTISIA DENTATA ASSEGAAIWOOD, ASSEGAAIHOUT

DISTRIBUTION: Is found in the Cape forests, through Natal to the moister districts of the Transvaal.

DESCRIPTION: This is a tall shapely, densely-leafed evergreen tree with a rounded crown, growing up to 18 m (60 ft) in height. The glossy leaves are a rich dark green on the upperside and paler underneath. They are broadly oval ending in a point and toothed, and measure 6-9 cm in length and 3-5 cm across. Young leaves and shoots are covered with reddish hairs giving them a bronzed and velvety appearance. The flowers are inconspicuous but the tree is worth growing because of its fine foliage and form. It is quick-growing and tends to become dense and bushy and it would therefore make a useful hedge or wind-break. It was once rated fourth amongst our valuable timber trees but it has been so heavily exploited that good specimens can now be found only in the most inaccessible places.

CULTURE: This is a good tree for a temperate climate, although once it has matured it will stand moderate frost and fairly long periods without water.

CUSSONIA CABBAGE TREE, KIEPERSOL

DISTRIBUTION: Can be found in the eastern Cape, Natal, the Orange Free State, the Transvaal and Rhodesia. The common name is derived from Kippe-solis, a corruption of a Portuguese word used in India for a paper parasol. It describes the arrangement of the leaves at the end of the trunk.

DESCRIPTION: These are decorative small trees of unusual form which can be used as accent plants in the garden. Cussonia sends up a rough bole which bears rounded whorls of leaves high up, giving the tree a rather palm-like appearance. The leaves are large, sage or green, and attractively indented. It is grown for its unusual shape and decorative foliage rather than for the flowers which it bears. Cussonia is most decorative as a pot or tub plant on a terrace or patio.

CULTURE: Cussonias do best in gardens where frosts are not severe. Once established they will survive long periods of drought and should be tried in regions which suffer from a low rainfall. Under good garden conditions and in a mild climate they grow quickly.

C. paniculata CABBAGE TREE, KIEPERSOL

This species grows in parts of the Karoo and other dry areas. It is seldom taller than 4-5 m (12-16 ft)

◁ The Cabbage Tree or Kiepersol *(Cussonia paniculata)* makes a fine accent tree.

and carries its clusters of leaves at the top rather like a feather duster. Each leaf is 30 to 60 cm long and made up of deeply indented leaflets. It makes a good accent plant and as it tolerates frost and long periods with little water it should be planted more freely in gardens where dry conditions limit the number of plants which can be grown. It bears woody branches of flowers.

C. spicata CABBAGE TREE, UMBRELLA TREE, KIEPERSOL

This species varies in appearance according to its habitat. Where rainfall is good it may reach a height of 10 m (30-35 ft), but in drought-stricken areas it seldom grows taller than 5 m (16 ft). It is a decorative plant similar to the species described above, but with darker leaves more deeply divided and larger spikes of flowers. The bark is thick and corky. It is not as slow-growing as *C. paniculata* and it is not as resistant to frost. It has large succulent roots which are eaten by some native peoples in time of drought and used by others as medicine for the treatment of malaria. The leaves are reputed to be good fodder for stock.

DAIS COTINIFOLIA DAIS, POMPON TREE, KANNABAS

DISTRIBUTION: Its natural habitat is the eastern part of the country, from the eastern Cape through the Transkei to Natal and the eastern Transvaal.

DESCRIPTION: This is one of the most attractive of our small trees and deserves to be more widely planted wherever conditions are suitable. It grows to about 5 m (16 ft) and has a rounded top of oval, bluish-green leaves with clearly defined midribs. In cold areas it is deciduous but in warm districts it is evergreen, as a rule. The leaves are approximately 3-6 cm long and 2-4 cm wide and more or less oval in form. In late spring it produces masses of round heads of pinky-lilac flowers. Each one is made up of a thin curved tube opening to five slender petals. The pompons of flowers are 5 cm (2 in) across and make a splendid show. It is a charming tree for small or large gardens and for street planting, and it is surprising that it is so seldom seen in gardens in Southern Africa. It was grown in Europe nearly 200 years ago, and it is recorded that one nursery in Australia has sold more than 6000 seedlings of this tree for gardens there, where it is extensively planted in private gardens and public parks. The bark was used by some Bantu tribes for tying down thatch on their huts, and the Voortrekkers used it for tanning leather.

Dais or Pompon Tree *(Dais continifolia)* showing detail of flowers and leaves.

Dais, Pompon Tree *(Dais continifolia)* showing tree in full flower.

CULTURE: Although its natural habitat is the warmer parts of the country this little tree will survive fairly severe frost and can be grown in highveld gardens provided it is given some protection during its first year or two. Once established it is able to stand long periods of drought. To encourage quick growth plant it in a hole filled with compost and manure, and water it well for the first two or three years.

DIOSPYROS TRANSVAAL EBONY, BLACK BARK

DISTRIBUTION: Diospyros occur in all parts of Southern Africa.

DESCRIPTION: The different species vary considerably in growth and according to habitat. Two species worth trying in the garden are described below.

CULTURE: The species mentioned are not quick-growing trees although, where winters are mild and they have an abundance of water, they are fairly fast. When once established they stand moderate frost.

D. mespiliformis TRANSVAAL OR RHODESIAN EBONY

Under warm humid conditions this tree will grow to a height of 18 m (60 ft). Under average garden conditions it grows to about half of this. Although it loses its leaves the old ones persist until the new ones form so that it is never completely bare of foliage. The dark green leaves are lance-shaped, about 10 cm long and 3 cm broad. The flowers are inconspicuous and are followed by small rounded fruits.

D. whytei BLACK BARK,
(*Royena lucida*) SWARTBAS

This is a handsome tree which, under optimum conditions, will reach a height of 12 m (40 ft). The oval leaves are small—seldom more than 2-3 cm in length and 1 cm across, but they make a dense and handsome crown of foliage as, when mature, they are dark green and glossy. Planted close together this species will make a good screen or windbreak.

DODONAEA VISCOSA SAND OLIVE, SANDOLYF, YSTERHOUT

DISTRIBUTION: Occurs in dry areas of the eastern Cape, the Orange Free State, the Karoo and the Transvaal.

DESCRIPTION: This is a small evergreen tree or large shrub which grows to 5 m (16 ft). The leaves are long and slender giving it a rather dainty appearance. The flowers are insignificant but the seed-pods are decorative. They are carried in clusters, are shiny and papery in texture, and coloured green, brown and burgundy. An infusion of the roots was used by Hottentots in the treatment of fevers. *D. thunbergiana* is a smaller species very similar in appearance. Both species make a good hedge or low windbreak, and both of them are useful for planting along roads to relieve the monotony of the landscape in arid areas, where few other trees or shrubs will grow.

CULTURE: These are not showy plants but they are useful ones for districts where aridity and frost make it difficult to grow a large variety of plants.

DOMBEYA ROTUNDIFOLIA WILD PEAR, WILD PLUM, DOMBEYA, DIKBAS

DISTRIBUTION: Is to be found in parts of Natal, the warmer areas of the Transvaal, South West Africa, Rhodesia and Malawi.

DESCRIPTION: This is a small deciduous tree growing to 5 m (16 ft) or a little more, with a neat crown of rounded leaves with serrated margins. The bole is covered with dark rough bark. The flowers appear in spring usually before the new leaves. They are white and make a splendid contrast to the dark branches. In winter when it is in blossom this tree is spectacular. The flowers, which look like the blossom of a pear, are carried in clusters and they appear in such numbers that the whole tree is wreathed in a white cloud of scented blossoms. It is one of the earliest of trees to flower before winter is over. The Swazi name "umBikanyaka" means "the tree that heralds the new season". Planted next to the kaffirboom, which flowers at the same time, it makes a brilliant show, as its white flowers enhance the brilliance of the scarlet of the kaffirboom flowers.

CULTURE: Although it grows best in warm regions it can be grown on the highveld, provided it is planted where it is protected, e.g. against a north-facing wall or near other trees. Once established it endures long periods without water.

EKEBERGIA CAPENSIS CAPE ASH, ESSENHOUT, MOUNTAIN ASH, DOG PLUM

DISTRIBUTION: Occurs from the south-west Cape to Natal, the Transvaal and as far as Ethiopia, usually in forests.

Wild Pear *(Dombeya rotundifolia)* is a beautiful sight in late winter.

DESCRIPTION: This tree is variable in growth. Under optimum conditions near the coast where winters are mild and there is abundant moisture and good soil it will reach a height of 10 m (30-35 ft) and more, whilst in arid areas or where winters are cold, it may remain stunted. It makes a good garden or avenue tree. It is evergreen in warm districts and semi-deciduous where winters are cold. The leaves are divided into glossy, pointed leaflets arranged in pairs with a terminal one. The creamy-yellow flowers which appear in summer are not showy but they produce a rather sweet scent like that of orange blossom which adds to the charm of the tree. The fruit is a berry about 18 mm in diameter which becomes crimson when ripe. *E. meyeri* (Mountain Ash) is a species which grows wild in the Transvaal. It is a handsome evergreen which grows well where winters are not extreme. Both species are worth a place in gardens and should be used more extensively also as street or avenue trees.

CULTURE: The dog plums are best planted in areas where winters are not severe. Plant them in holes filled with compost and manure, and water them abundantly for the first three or four years to promote robust growth. Once established they will stand long periods without water.

ERYTHRINA ERYTHRINA, KAFFIRBOOM

DISTRIBUTION: Occur from the eastern Cape, north through Natal, Swaziland and the Transvaal, to Botswana and Rhodesia.

DESCRIPTION: There are few trees from any part of the world which make as spectacular a show as our native erythrinas when they flower in late winter whilst the tree is still bare of leaves. Some species grow no bigger than shrub size and these are described in the section dealing with shrubs. The name is derived from a Greek word *erythros* meaning "red".

CULTURE: The kaffirboom can stand long periods of drought and moderate frost, when once it is established, but it does not do well in districts which have severe frosts or where winters are very damp. In the warmer parts of the country it is a quick-growing tree which needs little attention other than regular watering during its first year or two, to speed up growth. If planted at the beginning of the rainy season it will thrive even if not watered during its first dry season. It grows easily from truncheons, i.e. large pieces of stem. In some areas trees are attacked by borers but as yet no way of combating their attacks has been discovered.

E. caffra KAFFIRBOOM, LUCKY BEAN TREE

This species grows in the eastern Cape and north into the Transkei always fairly near the coast. In forests it may rise to 18 m (60 ft), but when planted in the open it is lower in growth with a more rounded crown. In the garden it grows to approximately 10 m (30-35 ft) with a spread of as much. It makes a good shade or avenue tree and it is surprising that it is not seen more often in gardens, or lining the streets of our towns. The tree has a grey trunk and branches which show up the luminous tomato-red colour of the flowers very well. The leaves, which start shooting out whilst the tree is still in flower, are divided into three leaflets, broad at the base and pointed at the apex. Both branches and stems are armed with thorns. The large bright spikes of flowers against a blue sky are a magnificent sight, and when the flowers fade they are followed by dark pods containing bright red seeds each marked with a black spot. In days past the flowering of the tree indicated to Bantu people that it was time to sow their crops.

E. latissima BROAD-LEAF KAFFIRBOOM
(*E. abyssinica*)

This species, which grows in warmer parts of Natal, the eastern Transvaal and Rhodesia, has smaller and less spectacular flowers than the species described above. It grows to about 6 m (20 ft), and has large, rounded leaflets which measure up to 20 cm across, and are sparsely covered with hairs.

E. lysistemon KAFFIRBOOM, LUCKY BEAN TREE

It is difficult to tell the difference between this and *E. caffra*. This species occurs naturally in certain districts in the eastern Cape, north into Natal and the warmer parts of the Transvaal and Rhodesia. The leaves are similar to those of *E. caffra* and so are the thorns. The flowers differ slightly in colour, and *E. lysistemon* has more slender flowers which do not have the protruding stamens that characterise *E. caffra*.

EUCLEA CRISPA GWARRIE

DISTRIBUTION: Is found in dry areas of the country.

DESCRIPTION: This is a useful little evergreen tree for dry areas where the number of plants which can be grown is limited by a shortage of water. It grows to about 5-7 m (15-22 ft), and has a rounded top of leathery, grey-green, oval leaves which vary considerably in size and shape. Male and female flowers are carried on different plants but neither of them is showy. The male flower is bell-shaped, cream in colour and sweetly-scented. The inconspicuous female flower develops into a small brown berry. The root of the plant is used by some African tribes to make a medicine for the treatment of rheumatism and as a purgative.

CULTURE: Gwarri will stand any amount of drought and a good deal of frost and is therefore to be recommended for planting on farms where shelterbelts are required for stock. It makes a good windbreak also. Where conditions are good it grows fairly quickly. It does not transplant well and should be sown where it is to grow to maturity, or else in peat-pots, which can be transferred to the ground when the plants are still small.

FAGARA DAVYI KNOBWOOD, KNOPDORING

DISTRIBUTION: May be found in forests in different parts of the country from the coast to the Transvaal.

DESCRIPTION: Although this is a tall tree under forest conditions it is unlikely to grow to more than 12 m (40 ft) in the garden under optimum conditions. The trunk of the tree is armed with knobs ending in sharp points. The leaves vary in size but they are generally large—up to 40 cm in length and made up of three to six pairs of leaflets with a terminal one. The leaflets are leathery, dark green and somewhat shiny, and have crenated margins.

CULTURE: Although this tree may stand fairly long periods of drought when once it is established, it is not recommended for areas with a very low rainfall. It will tolerate moderate frost but does best where winters are mild.

FAUREA TERBLANS, BOEKENHOUT

DISTRIBUTION: One species described has a limited distribution whilst the other occurs in many parts of Africa.

DESCRIPTION: The shape and size of these trees varies considerably from place to place according to growing conditions.

CULTURE: To ensure good growth plant them in well-prepared soil and water them regularly and well for their first four to five years. In areas of high rainfall they are fairly fast in growth. They are not suited to very dry areas or areas where frosts are severe.

F. macnaughtonii TERBLANS

This tree has a sporadic distribution; it is found in a few small stands widely separated—near George in the Cape, in one small area of Zululand and in another in the eastern Transvaal. Under good growing conditions the tree is large and upright with a pyramidal head. It grows to about

18 m (60 ft) in forests but in the garden it is likely to grow to only 12 m (40 ft). It has lance-shaped leaves 10-15 cm long with undulating margins. They are dark green and somewhat glossy on the upper surface. Young shoots and the stalks and veins of young leaves are crimson. In autumn it bears cylindrical spikes of scented white flowers 15 cm in length which hang down from the stems. This species is suitable for regions with a good rainfall and mild winters.

F. saligna AFRICAN BEECH, BOEKENHOUT
This species grows in the warmer parts of the Transvaal, in Natal, Zululand, Botswana, Rhodesia and further north. It reaches a height of about 9 m (30 ft), although occasionally much taller specimens may be found. It has a narrow and graceful crown of slender, leathery drooping leaves 5-10 cm long, which turn a pleasing shade of red in autumn. The flowers which are carried in cylindrical spikes are not showy. They are creamy-white to greenish-yellow in colour and have a honey-like scent. The bark of this species is very dark, rough and deeply fissured. *F. galpinii* is similar in appearance to this species. Both of them will tolerate drought and some frost when established.

FICUS WILD FIG
DISTRIBUTION: Different species are to be found growing wild in different parts of Southern Africa.

DESCRIPTION: Indigenous fig trees vary considerably in size according to species. The shapes of the leaves vary too. Some of them are too big for the ordinary garden but are good trees to plant in large gardens and on farms. They are useful as shade trees for stock. The missionary-explorer, Moffat, describes how he came upon a huge tree in the western Transvaal, where seventeen small huts were perched in the tree. The inhabitants had built their homes in the tree to be safe from lions and other animals. The leaves of the fig are oval with smooth margins and the stems exude a milky sap when broken.

CULTURE: Our wild figs do best in regions which do not experience very severe frosts and where the rainfall is 500 mm or more a year. Once established they can, however, stand long periods of drought.

F. capensis CAPE FIG, WILD FIG
In a forest this may reach a height of 21 m (70 ft) but in the open it seldom grows to more than 12 m (40 ft) in height. It is a spreading deciduous or evergreen tree with a thick bole and spreading roots, and it should therefore not be planted near buildings or drains. The leaves are broadly oval with pointed tips and may grow to as much as 15 cm in length and 10 cm in breadth. The margins of the leaves are sometimes smooth but often wavy. The fruits are round, approximately 2-3 cm in diameter, but not good for eating. This species, which grows from the coast to central Africa, makes a fine shade or avenue tree.

F. ingens RED-LEAFED FIG, ROOIWILDEVY
This species sometimes grows as a small tree and sometimes as a shrub scrambling over rocks and banks. In warm regions with a good rainfall it may grow to 9 m (30 ft). It is worth growing for its young spring foliage which is glossy and russet to red. It is hardy to moderate frost but in gardens which have severe frosts it should be planted where it will have some protection.

F. natalensis NATAL FIG, WILD FIG
Is a huge tree with glossy, dark-green leaves with rounded tips. It is handsome and quick-growing and worth a place on large estates to relieve the monotony of the landscape. Although it flourishes best in sub-tropical regions it also stands a certain amount of frost and drought, and has been grown in the Karoo.

F. pretoriae WONDERBOOM
A spreading evergreen species which reaches a height of 12 m (40 ft), or more, under optimum conditions, but which is usually much smaller than this. It is a good background tree for a large garden. The famous tree at Wonderboom outside of Pretoria is this species. The branches of the original tree bent down and took root and so new trees have developed about the first one. The spread of the Wonderboom there is now about 50 m and the height is circa 20 m. The parent trunk is thought to be 1000 years old.

FREYLINIA LANCEOLATA FREYLINIA
DISTRIBUTION: This tree occurs naturally from the south-western Cape towards Humansdorp.

DESCRIPTION: It is a small evergreen tree which reaches a height of 5 m (16 ft). It is upright in growth with drooping branches bearing long, slender leaves. In spring it carries clusters of sweetly-scented flowers of an orange-yellow shade. Each flower is tubular in form opening to five segments.

CULTURE: This is a quick-growing little tree which stands moderate frosts. It does not make a bright show but is a useful background plant.

GARDENIA GARDENIA, KATJIEPIERING

DISTRIBUTION: Gardenias grow wild in the eastern Cape, the Transkei, the warmer parts of Natal and the Transvaal, and Rhodesia.

DESCRIPTION: Most of the gardenias at present grown in our gardens are exotic plants, but our native ones are very attractive and well deserve a place in gardens where conditions suit them. They are small trees or large shrubs with sweetly-scented white or ivory flowers and neat attractive foliage.

CULTURE: Plant them in soil which is well enriched with compost and manure, and water them regularly, particularly during dry periods of the year. They will stand some drought, but they are very slow-growing plants and copious watering speeds up their rate of growth. They tolerate moderate frost but do best in gardens where winters are mild.

G. spatulifolia TRANSVAAL GARDENIA

This attractive species, which grows to 5 m (16 ft) is to be found in the lowveld of the eastern Transvaal, parts of Natal, Swaziland, Rhodesia and Botswana. Its leaves are about 4 cm long with a rounded apex and are narrower at the base than at the apex. The strongly-scented flowers consist of a slender tube which ends in an open starry face of waxy-white petals measuring 5 cm across. The flowers turn yellow as they age and are followed by large, roundish grey fruits. The flowering time is late spring and early summer. This species can survive long periods of drought.

G. thunbergia WILD GARDENIA, WILDE KATJIEPIERING

This species occurs in the eastern Cape and north into Natal and the warmer parts of the Transvaal. It reaches a height of 2-3 m (6-10 ft) and has leaves which are variable in size. They are glossy and rounded in the middle but pointed at the apex and the base and are decorative throughout the year. In summer it bears exceptionally handsome flowers with a rich scent which pervades the whole garden. The long slender buds look like a furled umbrella and the flower opens to a starry face at the end of a long tube. The flowers are followed by large, grey, woody, oval fruits, which hang on the tree for a long time. It grows from seed or truncheons—i.e. large sections of the stem. It is interesting to know that this species was cultivated under glass at Kew Gardens in England, nearly two hundred years ago.

GONIOMA KAMASSI KAMASSIE, KNYSNA BOXWOOD

DISTRIBUTION: This is a fairly common tree in the

A single plant of Wild Gardenia *(Gardenia thunbergia)* will scent the entire garden.

coastal forest near George, and some specimens are reported also to be growing in a forest in Zululand.

DESCRIPTION: It is a small tree seldom growing to more than 6 m (20 ft), and it has an attractive rounded crown of pretty foliage. The leaves are lance-shaped, dark green on the upper surface and paler underneath. They are glossy, 5-10 cm in length with a distinct midrib, and are carried in whorls of three or four. In late spring and summer the tree bears its tiny waxy-white flowers in small clusters at the ends of stems. Each flower is tubular ending in a starry face of five segments. They are not showy but they give off a strong scent rather like that of hyacinths. This is a handsome little tree for the small garden or patio.

CULTURE: Kamassie is rather slow-growing but it is attractive even when quite small. It needs good soil and abundant moisture and it will not endure very severe frost. It should not be tried in gardens where the rainfall is below 500 mm a year unless it can be watered abundantly.

GREYIA SUTHERLANDII NATAL BOTTLEBRUSH, MOUNTAIN BOTTLEBRUSH, BAAKHOUT

DISTRIBUTION: This is a small tree or large shrub which occurs in the eastern Cape, Natal and the eastern Transvaal.

Natal Bottlebrush *(Greyia sutherlandii).* Close-up of flower.

Natal Bottlebrush in full flower.

Mountain Bottlebrush *(Greyia flanaganii)* has attractive flowers in late winter.

Greyia radlkoferi is another showy species. It bears its flowers in spring.

DESCRIPTION: It reaches a height of 2-3 m (6-10 ft), and has leaves which are rounded and more or less heart-shaped, with serrated edges. The flowers which appear in spring before, or at the same time as the leaves, are cup-shaped with prominent stamens. They are of a rich, luminous scarlet colour and carried in showy spikes which make a brilliant show against the blue of the sky. Two other species worth growing are *G. flanaganii*, which is confined to the eastern Cape, and *G. radlkoferi* which occurs in the eastern and northern Transvaal. Their flowers are similar but they are not quite as spectacular when in flower as *G. sutherlandii*. These three make a splendid show in spring when they are in full flower.

CULTURE: Once established they endure long periods of drought, but only *G. sutherlandii* tolerates sharp frost. The other two species mentioned do best where winters are mild. They can be grown from seed, cuttings or suckers. After flowering cut off the faded blooms with a length of stem to keep the plants from becoming straggly.

HALLERIA LUCIDA TREE FUCHSIA, WITOLYF, NOTSUNG

DISTRIBUTION: Is found along streams and in forests in many parts of the country.

DESCRIPTION: It is an evergreen tree which usually grows to 5 m (16 ft), although in forests it sometimes reaches a height of 10 m (30-35 ft). The leaves are broad at the base tapering towards the tip, 3-5 cm long with serrated margins. The tubular flowers are not showy as they are partially hidden by the leaves. They are apricot to orange in colour and grow in small clusters close to the stem. When they fade berry-like fruits develop.

CULTURE: Halleria is a quick-growing little tree of no particular merit except to form a quick screen. It stands moderate frost when once established.

HARPEPHYLLUM CAFFRUM KAFFIRPLUM

DISTRIBUTION: Occurs in the eastern Cape, Natal and the warmer parts of the Transvaal.

DESCRIPTION: This is a pleasing evergreen tree which, under ideal conditions, will grow to 10 m (30-35 ft) or more. It has handsome foliage made up of four to eight opposite pairs of shining oval leaflets, 5-7 cm long, with pointed tips. They are dark green on the upperside and paler on the underside with a prominent midrib showing on both sides. The flowers are not showy but they are followed by plum-like oval fruits which turn scarlet when ripe. The fruit can be eaten raw or cooked to produce an acid jelly. It is a tree for the large garden or park, or for street or avenue planting. It can also be kept cut back to make a good dense screen or windbreak.

CULTURE: The kaffirplum grows fairly quickly and although it does best where winters are mild it can be grown on the highveld if given protection. It grows readily from seed or truncheons—i.e. large thick pieces of stem 30-90 cm long.

HOLARRHENA FEBRIFUGA JASMINE TREE

DISTRIBUTION: Can be found in warm districts of Rhodesia and Malawi.

DESCRIPTION: This is a decorative, small, deciduous tree for the small garden or patio. Sometimes it grows to only 3 m (10 ft) but, under good conditions it will reach a height of double this. The leaves are broadly oval ending in a sharp point, and measure 5-15 cm in length and 3-7 cm in breadth. They are green and glossy on the upper surface and greyish-green on the under surface. The margins of the leaves are often wavy. In mid-spring to summer the tree bears attractive clusters of white flowers which show up beautifully against the dark green of the leaves. The flowers give off a sweet scent rather like that of jasmine, which they somewhat resemble in appearance, too.

CULTURE: This is a slow-growing tree which does best in regions where frosts are not severe. Although it tolerates dry growing conditions it does better when watered regularly and well.

ILEX MITIS CAPE HOLLY, WATERBOOM
(I. capensis)

DISTRIBUTION: Is to be found from the Cape Peninsula through the eastern Cape to Natal and north into the Transvaal.

DESCRIPTION: This tree is variable in growth. In a forest it may reach a height of 15 m (50 ft) whereas in the open it seldom grows to more than 5 m

Red Mahogany *(Khaya nyasica)* has a pleasing form and foliage.

(16 ft). It is an evergreen with a rounded crown, and somewhat leathery, shining dark-green leaves up to 10 cm long and 2-3 cm wide with sharply defined midribs. They are paler on the underside than the top and bear some resemblance to the English holly. The white flowers are not showy but they are sweetly scented. They appear in summer in dense clusters and are followed, in late autumn and winter, by bright red berries relished by birds. The tree has a pale grey bark.

CULTURE: The Cape holly does best in regions where it receives an abundance of water. It is not a tree for dry parts of the country. Under good garden conditions it is fairly quick-growing. It survives moderate frost.

KHAYA NYASICA — RED MAHOGANY

DISTRIBUTION: Its natural habitat is Rhodesia, Malawi and further north.

DESCRIPTION: Under good growing conditions in forests of Malawi this tree may reach a height of 60 m (200 ft), but in the open it seldom grows to more than 18 m (60 ft), and in gardens where winters are not mild it may not grow to more than half of this. The bark is smooth and light brown and the tree has a fine, rounded crown of densely arranged leaves. The leaves are compound, made up of six pairs of broad, oval leaflets measuring 4-18 cm in length and 3-8 cm across. They are dark green and shining on the upper surface and paler on the underside. The white flowers which appear in spring are inconspicuous but sweetly scented. They are followed by large globular fruits.

CULTURE: This handsome tree is worth growing for its fine form and handsome foliage. It will stand some frost when established but does best in regions which have mild winters and a good rainfall.

KIGELIA AFRICANA — SAUSAGE TREE, CUCUMBER TREE
(K. pinnata)

DISTRIBUTION: Grows wild in warm districts of the Transvaal. Swaziland, Natal, Rhodesia and Malawi.

DESCRIPTION: Although it may reach a height of 18 m (60 ft), it seldom grows to this size. It is usually only 12 m (30-40 ft) tall and has a well rounded crown of handsome dark-green leaves 60 cm long divided into 3 to 5 pairs of large oval leaflets with a terminal one. The flowers are large and have crinkled petals opening up from a flared tube. The main flowering time is winter and early spring, when the tree is decorated with its sprays of crimson flowers which are full of nectar and attract many birds. When the flowers fade the tree becomes festooned with enormous cylindrical fruits which hang down on long stalks. They may measure up to 40 cm in length and 10 cm (4 in) across. Some African tribes use the powdered fruit to draw out ulcers. This is a tree only for the large garden or park, to plant in avenues, or to provide shade on farms.

CULTURE: The sausage tree does best in areas where frosts are never severe. It tolerates dry conditions but does best where the rainfall is not less than 500 mm a year. It is not suitable for very dry or very cold areas.

KIGGELARIA AFRICANA — WILD PEACH, WILDEPERSKE

DISTRIBUTION: Grows all over South Africa.

DESCRIPTION: This tree is very variable in size, being no more than a small scrubby tree in dry parts of the Karoo and a tree to 15 m (50 ft) in forests with a good rainfall. Generally it will grow to about 10 m (30-35 ft) with a spread of as much. It is an evergreen with leaves rather like those of a peach, although sometimes they are much wider and more rounded in form. They may have smooth or serrated margins. The flowers are not showy but the tree is worth growing because of the foliage. Planted close together it will form a good windbreak or screen and it is also useful on farms for providing shade and shelter for stock.

CULTURE: It stands both drought and fairly severe frost and is fairly quick, particularly if planted in good soil and watered regularly during the first three to four years.

KIRKIA ACUMINATA — WHITE SERINGA, WITSERING

DISTRIBUTION: This tree occurs in warm parts of the Transvaal, Botswana and Rhodesia.

DESCRIPTION: It is a graceful, deciduous tree which grows to approximately 10 m (30-35 ft) or more, with a spread of as much. It has smooth, pale grey bark when young and attractive leaves neatly divided into numerous little leaflets with serrated edges. The greenish-white flowers which appear in spring are inconspicuous, but the foliage turns attractive colours in autumn. Its new spring foliage is colourful too. It is an attractive tree for the large garden, park or avenue.

Silver Tree *(Leucadendron argenteum)* showing female flower.

The male flower of the Silver Tree *(Leucadendron argenteum)* is handsome and sweetly scented

The Silver Tree is worth growing for its decorative foliage.

K. wilmsii (the mountain seringa or wild peppertree) is a smaller species. It has foliage divided into small leaflets like the peppertree, which become most attractively tinted in autumn. Both species should be more widely grown.

CULTURE: These trees will stand long periods of drought but they are not hardy to much frost and, if being planted in regions where sharp frosts are experienced, they should be planted in a sheltered position. Where the soil is good and when watered well they are fairly quick-growing. They can be grown from seed or truncheons, i.e. thick stem cuttings.

LANNEA DISCOLOR LIVE LONG, BAKHOUT

DISTRIBUTION: Can be found in the lowveld and bushveld of the Transvaal, Rhodesia, Botswana, Malawi and further north.

DESCRIPTION: This tree varies in growth from place to place. It may remain a tall shrub to 3 m (10 ft) or it may grow to 9 m (30 ft) in height. It is deciduous with three to four pairs of opposite leaflets and a terminal one. The leaflets are oval, measuring about 4 cm in length. When young they are prettily covered with pink hairs and when mature the top surface is dark green and the underside is felted white. The flowers which appear in winter are not showy. They are followed by small fruits which turn deep purple when ripe and which are relished by birds. Planted close together this plant makes a good screen.

CULTURE: This little tree roots readily from large stem cuttings. In fact the common name of Live Long is said to be derived from the fact that poles used for fencing tend to take root and grow. It is a useful tree for dry gardens in regions where winters are mild.

LEUCADENDRON ARGENTEUM SILVER TREE

DISTRIBUTION: This tree is limited in its distribution to a small part of the south-western Cape near Cape Town.

DESCRIPTION: It is one of our most beautiful trees and worth trying in all parts of the country. Where climate and soil inhibit its development into a tree, it will nevertheless make an impressive show as a shrub. It is conical in shape when young and becomes broader at the top as it grows older. Mature specimens under optimum conditions may reach a height of about 10 m (30-35 ft), but generally it remains much smaller than this. It is a good tree to use as an accent plant in the garden, or to plant in large tubs on a terrace or patio. It was grown under glass in England almost 300 years ago. It is quick-growing if conditions suit it.

Although the flowers of this plant are attractive it is grown principally because of the beauty of its lance-shaped leaves, which are 6-15 cm long and covered with silver hairs which glisten in the sunlight, making the leaves look like pieces of highly burnished silver. These hairs are nature's protection against excessive drying out from the winds which sweep the slopes of its natural habitat. In winter, or during rainy weather, the hairs are raised a little to allow air to circulate, whilst in dry weather they press down closely against the leaves to reduce transpiration.

The male and female flowers are carried on separate trees. The male flower is the more handsome of the two. It is made up of a rounded head coloured from ivory to crimson and grey and is surrounded by a halo of silvery leaves. It has the scent of a vanilla custard. The female flower forms a silvery cone. The flowering time is winter.

CULTURE: Silver trees should be planted in soil which is acid and they must be watered regularly during autumn and winter. To give them a good start make holes 1 m (3 ft) wide and deep, and fill these with acid compost and good soil. They do not mind being somewhat dry in summer and they stand moderate frost, provided they are watered. When being grown in areas which have frost the young plants should be protected.

LOXOSTYLIS ALATA — TIERHOUT, WILD PEPPER TREE

DISTRIBUTION: Can be found near the coast between Port Elizabeth and the Kei River.

DESCRIPTION: This tree does not bear showy flowers but it is worth growing because of its shining decorative foliage. It is an evergreen with divided leaves. The leaflets are somewhat leathery and arranged in pairs with an odd one at the end. It reaches a height of about 8 m (26 ft) with a spread of about 6 m.

CULTURE: Tierhout grows in any kind of soil and once established it will stand drought and moderate frost. To encourage quick initial growth water it well for the first three to four years.

MILLETTIA GRANDIS — UMZIMBEET, KAFFIR IRONWOOD
(Millettia caffra)

DISTRIBUTION: Grows in the coastal forests of the Transkei and Natal.

DESCRIPTION: This is a handsome tree with glossy, divided leaves, each leaflet being about 5-7 cm in length. It is generally an evergreen but in cold winters it may lose most of its leaves. The new foliage is tinged brownish-red and the tree is a fine sight throughout the year. In its natural habitat it grows into a very large tree, but in gardens it seldom grows to more than 10 m (30-35 ft) with a spread of about the same. In summer it has spiky sprays of purple, pea-shaped flowers each one about 2-3 cm long. These are followed in autumn by large, woody pods covered with brown hairs. This is a fine tree for shade or for avenues or street planting where conditions are suitable. There is another handsome species known as *M. sutherlandii,* which is larger and more suited to park-planting than for the garden.

CULTURE: Umzimbeet tolerates some frost but grows quickest in mild climates where it should

Umzimbeet *(Millettia grandis)* does best in warm regions.

be more widely planted for shade and shelter on farms as well as for beautifying gardens and streets. It is not suitable for areas where frosts are severe or where long periods of drought are common.

MIMUSOPS — MILKWOOD TREE

DISTRIBUTION: The three species described below are found from the eastern Cape through Natal to the Transvaal, and north into Rhodesia.

DESCRIPTION: They are slow-growing, evergreen trees which are variable in growth according to climatic conditions. They all have fruit relished by birds, and the stems when broken exude a sticky, milky sap. Where conditions are suitable they should be planted more extensively for shade and shelter on farms or as background trees in coastal and inland gardens.

CULTURE: These trees will not thrive in very dry areas and should be planted only where the rainfall is fairly good or where they can be watered.

In winter Vlier or White Elder *(Nuxia floribunda)* becomes festooned with large heads of ivory flowers.

M. caffra RED MILKWOOD, MOEPEL
Grows right down to the beach in the eastern Cape and Natal. It is not of attractive form but useful inasmuch as it tolerates sandy soil and salty air. This is a very slow-growing tree.

M. obovata RED MILKWOOD, MOEPEL
This species is to be found in the eastern Cape and north into the Transvaal. It grows to about 10 m (30 ft) and has a rounded head. The leaves are approximately 3-7 cm long, oval and rounded at the tips. Old leaves are dark green and have a sheen, whilst young ones are covered with down. The flowers are not showy and the oval, pointed fruits which follow are coloured orange when ripe. They have an acid flavour.

M. zeyheri TRANSVAAL RED MILKWOOD, MOEPEL
This species grows in the northern Transvaal and in Rhodesia. It reaches a height of approximately 9 m (30 ft) and has a rounded head. The leathery leaves are dark green and fairly glossy, about 7 cm long and 2-3 cm wide. Young twigs and the undersides of young leaves are covered with russet coloured hair. The flowers are fragrant but insignificant and are followed by yellow fruits.

NUXIA FLORIBUNDA VLIER, WHITE ELDER
(Lachnopylis floribunda)
DISTRIBUTION: Its natural habitat is mainly the forests of the southern and eastern Cape and Natal, but it is also found occasionally in the Transvaal.

DESCRIPTION: This is a decorative tree growing to about 9-12 m (30-40 ft) in height, with a rounded crown when mature. The oval, pointed leaves are approximately 20 cm long and 5 cm wide and fairly light in colour. The margins may be entire or slightly serrated. The leaves are arranged in whorls of three. In winter it is a glorious sight when covered with large clusters of tiny, ivory-white flowers, with the appearance of old lace. The clusters very often measure 30 cm (12 in) across. It is a fine tree for the garden, for parks and for lining streets and avenues.

CULTURE: Vlier is quick-growing in regions which have mild winters and a good rainfall. In areas where the air is dry for most of the year it will remain small but nevertheless decorative. It stands moderate frost. It can be grown from seeds or cuttings.

OCHNA PULCHRA OCHNA, LEKKERBREEK
DISTRIBUTION: Can be found in warm parts of the Transvaal, Rhodesia, Malawi and Botswana.

DESCRIPTION: This is a decorative small deciduous tree suitable for small and large gardens and for pavement planting. It reaches a height of about 6 m (20 ft) and has a narrow, rounded crown. The trunk is usually mottled pale grey as the bark tends to peel off. In spring the new foliage is beautiful—glossy and shaded from pale green, through copper to rose. The bright yellow, cup-shaped flowers appear in spring too, but they do not last for very long. After they fade the sepals of the flower enlarge and turn crimson making a colourful background to the fruits, which become jet black in summer, when ripe. The tree is showy for a long period. Some native tribes of Malawi carry a piece of the root as a lucky talisman. The name "Lekkerbreek" was given because the branches break easily. *Ochna arborea* (Cape Plane or Cape Redwood) is another small tree worth trying in gardens, but it is not as pretty as the species described. It grows wild in the coastal area from Knysna through the Transkei into Natal. The name "Redwood" describes the reddish colour of the wood. Both species are worth growing where climatic conditions are right.

CULTURE: These little trees stand moderate frost and fairly long periods of drought, but do best in areas with warm winters and a good rainfall. They are not quick-growing but well worth planting. The plants are not easy to propagate from seeds or cuttings.

OLEA AFRICANA WILD OLIVE, OLIENHOUT

DISTRIBUTION: Occurs in many parts of South Africa, Rhodesia, Botswana and South West Africa.

DESCRIPTION: This evergreen tree is not highly ornamental but it is a useful one for areas where growing conditions are difficult. It has somewhat leathery lance-shaped leaves from 5-8 cm long and about 15 mm wide. They may be green or grey-green in colour. Where the rainfall is good the plant grows into a better shape and has larger and more attractive leaves than in dry areas. Under optimum conditions it may grow to 10 m (30-35 ft) but in its natural habitat it is seldom more than 6 m high. The flowers are inconspicuous and the fruit which turns black when ripe is very small. The juice of the fruit was used by early settlers to make ink, and the leaves were used for tea. In dry areas of the Karoo the leaves are used as fodder for stock and the berries which ripen in March are relished by birds. The timber is very hard and lasts for years when used as fence poles. The wild olive is recommended for shelter belts for stock, and for garden and street planting in regions where frost and aridity makes it difficult to grow other trees. It is also a useful tree for close-planting to form a dense tall hedge or windbreak. An infusion made from the leaves is used by some African tribes as an eye lotion.

CULTURE: It is a slow-growing tree which should have the side shoots cut out to make a canopy of the top. Once established it stands both frost and drought and it does well in alkaline soil.

OLEA CAPENSIS subsp. **MACROCARPA** BLACK IRONWOOD, YSTERHOUT

DISTRIBUTION: This tree occurs mainly in forests of the Cape Province but it is found in the Transvaal too.

DESCRIPTION: Under forest conditions it may grow to about 18 m (60 ft) but it is unlikely to reach more than 10 m (30-35 ft) under garden conditions. It has a rounded head at the top of a long, straight grey bole. The leaves are slightly glossy, and leathery, darker green on the top surface than the underside, about 5-10 cm long and 2-4 cm wide, tapering at both ends. The cream flowers are not showy and they later develop into small oval fruits like an olive in shape. They are dark purple when mature and are relished by birds and animals. This is a handsome evergreen tree for the large garden.

CULTURE: It will stand moderate frosts and some drought when established, but it is not recommended for areas where the rainfall is less than 500 mm a year. Under good conditions it is fairly fast in growth.

OLINIA CYMOSA HARD PEAR, ROOIBESSIE

DISTRIBUTION: Occurs in the coastal forests of the Cape.

DESCRIPTION: This is the fastest-growing tree in the Knysna forests. It is evergreen, handsome and hardy. The glossy dark green leaves which are paler on the underside are obovate and sometimes pointed. They measure 3-7 cm in length and are closely arranged on the stem in pairs opposite each other. Leaves, twigs and stems when crushed smell of almonds. The tiny white flowers are arranged at the ends of stems in elongated clusters. The fruit which is the size of a pea is red when ripe. This tree, which grows to 24 m (80 ft) under forest conditions, may not reach more than 12 m (40 ft) in the garden. It is suitable for the large garden rather than the small one. *O. emarginata*, which occurs in the Transvaal, is a smaller species with more slender leaves. It is worth planting where a quick-growing evergreen tree is required. A third species *O radiata* is similar to *O. cymosa* but it bears larger fruits.

CULTURE: All species will stand fairly severe frost, and once established, they will also endure somewhat dry growing conditions, too. They do best, however, in areas where winters are not very severe.

PARINARI CURATELLIFOLIA MOBALA PLUM, GRYSAPPEL

DISTRIBUTION: Occurs in the Transvaal, Botswana, Rhodesia, Malawi and further north.

African Wattle *(Peltophorum africanum)* is a decorative shade tree.

DESCRIPTION: This is an evergreen tree with a rounded crown. Under average conditions it grows to 9 m (30 ft), although in Malawi, Mozambique and Zambia specimens of double this size are found. The leaves are oval, about 5-7 cm long and 2-4 cm broad, with clearly defined veins which run at right-angles from the midrib to the margins of the leaves. The creamy-white flowers, which appear in spring, are inconspicuous but sweetly scented. They are followed by plum-like fruits which become yellow when mature. Some native tribes brew a drink from the fruits. They are also very popular with many tribes, as fruit, or as a relish to be eaten with porridge. The roots of the tree are used in medicine and magic. It is a fine tree for the large garden or park and for planting along a street or avenue.

CULTURE: This tree will stand moderate frost but it is not recommended for cold gardens. It is slow-growing particularly in dry areas.

PELTOPHORUM AFRICANUM AFRICAN WATTLE, RHODESIAN BLACK WATTLE, HUILBOOM

DISTRIBUTION: Is to be seen in the warm areas of the Transvaal, Swaziland and Rhodesia.

DESCRIPTION: This is a decorative, deciduous tree with feathery foliage of greyish-green. The leaves are divided into very small leaflets which look rather like those of an acacia. It grows to 10 m (30-35 ft) with a spread of as much, and in spring it has masses of five-petalled, yellow flowers carried in spikes, which make a fine show. When they fade the tree becomes festooned with brown seed-pods about 5 cm long. The bark is rough and dark brown, and when the tree is bare of leaves it has a rather gaunt appearance. The name of Huilboom, meaning "weeping tree" is given because of the fact that in late spring water drips from the stems of some of these trees making pools on the soil beneath them. This is thought to be due to the activities of certain insects. It is an ornamental shade tree for the garden or park and also a good one to use along an avenue or drive.

CULTURE: Given good soil and an abundance of water this tree is fairly quick-growing, but although it tolerates some frost it is not suited to areas where temperatures drop very low in winter. Once established it will endure long periods of drought.

PITTOSPORUM VIRIDIFLORUM UMKWENKWE, KASUUR

DISTRIBUTION: Occurs in the eastern Cape and further north into the Transvaal and Rhodesia.

DESCRIPTION: This species is not as handsome as some of the exotic species of pittosporum but it is more tolerant of drought and frost. It is a large shrub or small evergreen tree which grows to a height of 6 m (20 ft). The leaves are 4-10 cm long and about 3-4 cm wide, broader at the apex than the base, and leathery in texture. They are glossy and dark green on the upper surface and pale green on the underside with a prominent mid rib. The leaves tend to be clustered together on the stems. In spring it bears small, greenish-yellow, scented flowers which are inconspicuous. These are followed by orange seeds.

CULTURE: This plant does better at the coast than inland and makes a useful windbreak in coastal gardens. It stands moderate frost, but is slow-growing unless watered fairly regularly. It is a useful little tree for difficult conditions but may seed or sucker too freely in the garden where growing conditions are good.

PODOCARPUS YELLOWWOOD

DISTRIBUTION: Yellowwoods are found from the southern Cape east towards Natal and north into the northern Transvaal.

DESCRIPTION: They are evergreen forest trees of huge size which belong to the group of trees known as conifers. Several species are to be found in South Africa some of which are described below. Generally they are to be considered as trees suitable for large estates or parks rather than for the ordinary garden, but they can also be grown as ornamental plants in pots or tubs on a patio. When their roots are restricted by a container they never grow to more than a few feet in height.

CULTURE: Yellowwoods are fairly slow-growing evergreen trees which thrive naturally in regions where the soil is good and where the rainfall is fairly high; they are unsuitable for dry hot areas, or for areas which experience long periods of drought and severe frost, but they endure cold winters where there is enough humidity.

P. falcatus OUTENIQUA OR COMMON YELLOWWOOD

Under forest conditions this tree reaches a height of approximately 50 m (160 ft) but when grown on its own it is unlikely to grow to more than about 12 m (40 ft) in height. It is upright with a rounded crown of narrow leaves of a pleasing shade of green.

P. henkelii HENKEL'S YELLOWWOOD

Can be found in the northern Transkei and southern Natal. This is the most decorative species for the large garden. In a mature forest it reaches a height of about 30 m (100 ft) but in the open it grows slowly to approximately 10 m (30-35 ft), and is thickly clothed with slender leaves arranged in drooping tufts right up the tree. It also makes a fine tub plant for the patio which is shaded for part of the day.

P. latifolius REAL YELLOWWOOD, UPRIGHT YELLOWWOOD

This species, which can be seen in the Knysna forests growing to 40 m (130 ft) or more, has a crown which is small in proportion to its stature. Under open conditions it will probably grow no taller than 12-15 m (40-50 ft). It is not as ornamental a tree as the species described above, but worth trying in large parks and gardens. This is the species which was extensively used for making the floors and ceilings of the old Cape homes.

PRUNUS AFRICANUM RED STINKWOOD,
(Pygeum africanum) BITTER ALMOND

DISTRIBUTION: Grows in forests throughout the country but only in small numbers.

DESCRIPTION: This is an evergreen tree which, under forest conditions, may reach a height of 20 m (65 ft), but which under garden conditions will probably not grow to more than 10 m (30-35 ft). It has a crown of dark green, glossy leaves which are oval in shape tapering to both ends and measuring 7-10 cm in length, with a sharply defined midrib. The flowers are not showy and the tree is grown chiefly for its shape and the foliage. The bark is dark and rough, and the little fruits which follow the flowers are about the size of a small cherry and unpleasant to taste. It is a useful tree for parks and large gardens and for an avenue. The bark is reddish brown and has an unpleasant odour which accounts for its common name.

CULTURE: This tree is moderately fast in growth when planted in areas where there is regular rain or if watered well in its early years. It is not suitable for gardens where the rainfall is low or where frosts are severe, but it will tolerate moderate frosts.

Although slow-growing the Yellowwood is attractive when young.

PSEUDOCADIA ZAMBESIACA NYALA

DISTRIBUTION: Can be found in the warmer parts of the Transvaal and Rhodesia.

DESCRIPTION: This is an evergreen tree which reaches a height of about 10-12 m (35-40 ft). It has a rounded head and makes a good shade tree for the large garden or park and for broad avenues. The leaves are divided into oval, pointed leaflets which are dark green on the upper surface and slightly glossy. In late spring and early summer the tree produces sprays of small white flowers. These are followed by oval, brown fruits which are relished by birds and buck.

CULTURE: It grows readily from seed and stands drought but is tender to frost.

PTEROCARPUS KIAAT, BLOODWOOD, TRANSVAAL TEAK

DISTRIBUTION: Occurs in the bushveld and lowveld of the Transvaal, and further north in Rhodesia, Zambia and Malawi.

DESCRIPTION: These trees are variable in growth according to climate. They have a rounded or fairly wide crown and attractive leaves. The flowers may be showy for a short period.

CULTURE: They do best in a warm climate where the rainfall is not less than 500 mm a year and where frosts are never more than moderate. They grow quickly from truncheons (long pieces of stem) planted in October.

P. angolensis KIAAT, TRANSVAAL TEAK

This is a fairly large deciduous tree which grows to about 9 m (30 ft) or more, with a fairly spreading top. In warm parts of the country where the rainfall is good, it may grow to much more than this. It has compound leaves divided into nine to twelve pairs of oval, pointed leaflets which are broad in proportion to their length. They measure 3-8 cm in length. The yellow to orange, scented, pea-shaped flowers appear in sprays at the ends of branches in spring, at about the same time as the new leaves unfurl. When the flowers fade the tree bears round winged seed-pods 7-10 cm in diameter.

P. rotundifolius ROUNDLEAF KIAAT, BLINKBLAARBOOM

This is a much smaller deciduous tree. It has shiny, rather pale-green roundish leaflets, and sprays of sweetly scented, golden-yellow flowers in late spring, followed by winged fruits. Given good soil and sufficient water it is a quick-growing tree.

RAPANEA MELANOPHLOEOS CAPE BEECH, BOEKENHOUT

DISTRIBUTION: Can be seen in the forests of the Cape, Natal, the Transvaal and further north.

DESCRIPTION: This tree is very variable in growth and may under difficult conditions be no larger than a shrub whilst on the edge of a forest it may grow to a height of 6-12 m (20-40 ft). It is an evergreen tree with a rounded crown of lance-shaped, leathery leaves which are dark green on the upperside and paler beneath. The midrib is rather prominent on the undersides of the leaves. They are 6-15 cm in length and have blunt ends. The flowers are insignificant but have a faint scent which attracts bees. Male and female flowers are carried on separate trees and the female ones are followed by berries of dark purple. When well grown this makes an attractive shade tree.

CULTURE: The Cape beech is a fairly quick-growing tree but it is inclined to send up suckers and it should not therefore be planted in a small garden. It is useful as a windbreak in areas where the rainfall is good.

RAUVOLFIA CAFFRA QUININE TREE, NCHONGO

DISTRIBUTION: Occurs in the eastern Cape, the Transkei, Natal and the Transvaal lowveld, and further north in Malawi and Kenya.

DESCRIPTION: In a warm climate this is an evergreen tree. It grows to 12 m (40 ft) in height, if planted in good soil and watered well. It does not have showy flowers or fruits but its foliage is attractive. The leaves are long and pointed at both ends and up to 15 cm in length. They are smooth and glossy and arranged in whorls of four. It is a very handsome tree for large gardens or parks, or for broad streets or avenues. At one time the bark was used in the treatment of malaria.

CULTURE: This is a slow-growing tree which stands only moderate frost, and which needs an abundance of moisture to encourage growth. It is not suitable for dry or cold gardens.

RHUS CHIRINDENSIS RED CURRANT, BOSTAAIBOS
(R. legatii)

DISTRIBUTION: Can be found in forests of the south-western Cape and the Transkei.

Description: Under optimum conditions in a forest it reaches a height of 24 m (80 ft) but in the open it is unlikely to grow to more than half this height, and often it may be much less. It is the tallest and most decorative species of this genus in South Africa. A well-grown tree has a rounded crown of soft green leaves made up of three oval pointed leaflets, each one 5-10 cm in length and 3-5 cm in breadth. The leaves are carried on long stalks giving the tree a graceful appearance. The flowers appear in spring in long loose trusses up to 25 cm in length, and in autumn the tree is adorned with bunches of dark red, shining berries which attract birds from far and wide. When the stem is cut a red sap is exuded. This is a deciduous tree worth growing where conditions are suitable as it is handsome in form and it makes a good shade tree. It tends to throw out coppice shoots which form a dense cluster of stems and it could therefore also be used as a windbreak in a large garden.

Culture: The red currant needs good soil and an abundance of water for its best development. It is not recommended for areas where the rainfall is less than 500 mm a year or where frosts are severe. Under good conditions it is quick-growing.

RHUS LANCEA — Karee, Karoo Tree, Bastard Willow

Distribution: Is to be found in many parts of the country and particularly in dry areas of the northern Cape, the Karoo, Transvaal, Orange Free State, South West Africa and Botswana.

Description: It varies in shape from a bush to a tree with a rounded crown about 8 m (26 ft) high, depending upon the growing conditions. The bark is usually rough and dark in colour. The leaflets are long and slender and arranged in threes, the middle one being longer than the other two. They droop down in a rather effective fashion. The under surface of the leaves is paler than the top surface and the midrib is clearly defined. The tiny flowers are insignificant and followed by clusters of small fruits which are fermented by some indigenous tribes to make beer. When mature it makes a good shade tree for the garden and it is also a good tree to use for street and roadside planting in dry areas. Planted in groves on farms it gives shade and shelter for stock, and closely planted, it makes a good windbreak. The leaves are said to be useful as fodder for stock. *R. leptodictya* (Bergkaree) and *R. viminalis* are two other native species very similar to *R. lancea*. Their leaves are broader.

Culture: The karee is a good evergreen tree to plant in dry parts of the country where winters are apt to be cold for it stands both drought and severe frost. It also grows well in alkaline soil. To encourage quick growth prepare holes with compost and manure before planting, and, if possible water the young trees well until they are established. They can be grown from seed or cuttings.

SALIX CAPENSIS — Cape Willow, Wild Willow

Distribution: Its natural habitat is along streams and rivers from the eastern Cape north into the Transvaal. It is common along the Orange and Vaal rivers.

Description: This tree varies in height and spread considerably depending upon climatic conditions. It may be a large bush or a tree to 10 m (30-35 ft). It has slender, lance-shaped leaves carried on branches which have a tendency to droop, but not as markedly as those of the weeping willow. The young leaves have smooth edges but older ones are serrated along the margin. The bark on the older branches is fairly dark and rough. Generally it is deciduous but it may occasionally retain its leaves through winter. The flowers which appear in spring are insignificant. The male ones are carried in catkins and the female ones in spikes. The leaves are eaten by stock and are used by some native tribes to make a remedy for rheumatism. The Cape willow should be more extensively planted as a background plant in regions where gardening is difficult. Other indigenous species to be found along streams are *S. hirsuta*, *S. mucronata*, *S. woodii* and *S. wilmsii*.

Culture: The Cape willow grows readily from large thick stem cuttings and, once established, it stands considerable frost and a good deal of drought too.

SCHOTIA — Boerboon, Hottentots Bean

Distribution: Different species are to be found in different parts of the country.

Description: The schotias vary considerably in size and shape. They deserve to be more widely planted not only to provide shade in gardens but also to give shelter and shade on farms. They produce large seed-pods enclosing bean-like seeds which were relished by Hottentots and other native people, and which the early colonists also found to be a nourishing addition to their sparse diet. This accounts for the common name of *boerboon* which means "farmer's bean".

CULTURE: They grow readily provided they are watered fairly regularly when young. Once established they will stand long periods of drought and they will also tolerate moderate frost, but they are rather slow-growing in a dry climate.

S. afra KAROO BOERBOON, HOTTENTOTS BEAN

This species is seen in different parts of the Karoo. It is more like a shrub than a tree and useful as a windbreak or tall hedge in dry areas of the country. It is slow-growing but resistant to drought. In spring when it bears its crimson flowers it is quite a gay sight, and later it bears colourful pods of shades of green, pink and russet-brown.

S. brachypetala WEEPING BOERBOON, HUILBOERBOON, TREE FUCHSIA

It is occasionally found in the eastern Cape but is more common in warm parts of the Transvaal, Natal and Rhodesia. It is an evergreen tree in warm climates but may lose its leaves when planted in areas where winters are cold. It grows to 10-12 m (30-40 ft), and has a fairly wide and rounded top. The leaves are made up of four to seven pairs of leaflets broad at the ends and about 4 cm in length. In spring the tree is very attractive as its new foliage is tinted in shades of red, yellow and copper. As the leaves mature they turn first pale green then dark green. It bears deep crimson flowers on the old wood. They have an abundance of nectar which attracts bees and birds. The flowers are followed by large brown pods up to 10 cm long.

S. latifolia FOREST BOERBOON

This species is limited in distribution to the area from George to the eastern Cape. It is very similar in form and foliage to *S. brachypetala*. Its pink flowers have distinct petals and are carried at the ends of stems whereas the flowers of *S. brachypetala* have shortened perianth segments and are to be found on the old wood. It grows to about 6 m (20 ft) in height.

SIDEROXYLON INERME WHITE MILKWOOD, SEA OAK

DISTRIBUTION: Occurs along the coast from the south-western Cape east into Natal and further north.

DESCRIPTION: This is an evergreen tree variable in growth from a large shrub to a tree of about 5 m (16 ft) with a spreading crown of leaves. The leaves are leathery, glossy and dark green on the top surface and paler beneath. They have a clearly defined midrib on the underside, and are broadly oval with rounded ends. The broken stems usually exude a milky latex. It bears small, insignificant, creamy-white flowers in clusters. They give off a smell which is sometimes rather unpleasant.

The Weeping Boerboom *(Schotia brachypetala)* grows well in warm areas.

Stock Rose *(Sparrmannia africana)* is a quick-growing tree with attractive foliage.

The flowers are followed by fruits which are small and purple.

CULTURE: This is a slow-growing tree suitable for coastal gardens as it stands salt-laden air and does not mind sandy soil.

SPARRMANNIA AFRICANA STOCK ROSE

DISTRIBUTION: Its natural habitat is the southern Cape between Knysna and Port Elizabeth.

DESCRIPTION: This is a very quick-growing small tree or large shrub. Often it sends up so much basal growth that it is more like a shrub than a tree, but, by cutting out some of these stems at the bottom, it can be trained to form a very pretty tree, growing to a height of about 6 m (20 ft). The leaves are very effective, and for this reason the plant has been used in England for many years as a pot plant. The leaves are heart-shaped and very large, measuring up to 20 cm in length, and are soft in texture, with serrated edges. The flowers

Stock Rose *(Sparrmannia africana)*. Close-up of flower.

which appear in spring are carried in rather pretty clusters. They are white and enhanced by their bright yellow stamens.

CULTURE: This is a good plant for small or large gardens in regions where frosts are not severe. It grows in any kind of soil and in full sun as well as in partial shade. In inland gardens it does better if planted where it is shaded for some hours of the day. When cut down by frost the plant usually recovers rapidly. It is advisable to water it regularly and well throughout the year until it is well established.

SPATHODEA CAMPANULATA AFRICAN OR RHODESIAN FLAME TREE

DISTRIBUTION: Occurs naturally in the warmer parts of Rhodesia and East Africa, and further north.

DESCRIPTION: This is one of the most handsome of the trees indigenous to Southern Africa. Under forest conditions it reaches a height of about 16 m (50 ft) or more and spreads across about 8-10 m, but in the garden it grows to a little more than half of this. The huge leaves are divided into large, oval leaflets measuring about 10 cm in length and 5 cm across. They are boldly veined. The tree is generally evergreen but may lose most of its leaves in cold regions. The flowers appear from late winter to summer. The large flower buds are curved and clustered together rather like a bunch of bananas, forming a flowerhead 18 cm across. They are brownish-green and open in twos and threes to reveal glorious, chalice-shaped flowers about 6 cm across. They are coral-red and attractively edged with gold.

CULTURE: Although tender to frost when young this tree stands quite sharp frost when established and it endures fairly long periods of drought also. Where it is watered regularly and well and where the soil is good it is fast-growing. It is not suitable for areas of severe frost or where the rainfall is less than 500 mm a year.

STERCULIA MUREX LOWVELD CHESTNUT

DISTRIBUTION: Occurs in the lowveld of the Transvaal and Swaziland.

DESCRIPTION: This is a small, deciduous tree with a rounded crown, growing to 9 m (30 ft). The bark is rough, deeply fissured and greyish-brown in colour. The dark green leaves are digitate, like those of the horse chestnut. Each leaf is made up of 5-7 narrow leaflets emerging like fingers from the palm of the hand. The yellowish flowers marked with brown lines are carried in attractive sprays in late spring, usually before the leaves appear. Each flower has five recurved waxy calyx lobes which look like petals. The flowers are followed by large, rounded seed-pods covered with spines. The tree is worth growing because of its pretty foliage.

CULTURE: This is a tree for gardens where winters are mild. It can endure long periods with little rain but is sensitive to frost.

SYZGIUM GERRARDII WATER PEAR,
(Eugenia gerrardi) FOREST WATERWOOD

DISTRIBUTION: Is limited to forests of Natal, the northern Transkei, parts of the northern Transvaal, and Rhodesia.

DESCRIPTION: This is a handsome evergreen tree which reaches a height of approximately 15 m (40-50 ft) or more under forest conditions, but only about 10 m (30-35 ft) in the open. It does not bear showy flowers but is worth growing as a shade tree in a large garden where the climate is suitable. *S. cordatum* is another native species which should be planted in more gardens and parks as it makes a fine shade tree.

CULTURE: They grow fairly rapidly where their roots can get down to water or where the rainfall is high. They are not recommended for gardens in dry areas or where frosts are severe.

TAMARIX AUSTRO-AFRICANA WILD TAMARISK, ABIEKWAS

DISTRIBUTION: Occurs along river beds in the Karoo, Namaqualand, South West Africa and the northern Cape.

DESCRIPTION: It grows into a large bush or small tree with greyish leaves rather like those of a cypress. Its slender stems give it a graceful appearance, and in spring it bears quite showy little pink flowers. The foliage is said to provide good fodder for stock.

CULTURE: This is a useful small tree for gardens where the soil is alkaline and sandy, and where the rainfall is low. It also tolerates fairly severe frost.

TARCHONANTHUS CAMPHORATUS
WILD SAGE, CAMPHOR WOOD, VAALBOS

DISTRIBUTION: This tree occurs in many parts of the country from the coast to dry areas inland, up into Botswana.

◁ The Rhodesian Flame Tree *(Spathodea campanulata)* has magnificent flowers and decorative foliage.

DESCRIPTION: This is not a handsome tree, but because of its tolerance of difficult growing conditions it is worth growing in areas where a wide variety of other trees cannot be grown. It sometimes is no larger than a shrub in size but under better conditions it reaches a height of 6 m (20 ft) and has a rounded crown. The leaves are leathery and lance-shaped, grey-green on the upper surface and paler below. Generally the margins are smooth but occasionally they are finely toothed. When bruised they give off a scent of camphor. The creamy-white flowers are not showy but the plant is worth growing for its foliage effect, and is recommended for coastal gardens and those in dry areas. In the sandveld of the Kalahari and parts of Botswana this plant provides fodder for the livestock.

CULTURE: It grows right down to the coast where it does not seem to mind salt spray and it tolerates extremes of drought, wind and cold. It grows easily from seed and could therefore be treated as a windbreak where wind conditions make gardening difficult. It could also be planted to arrest the drift of sand at the coast and inland.

TERMINALIA TRANSVAAL SILVER TREE, HARDEKOOLBOOM

DISTRIBUTION: The species mentioned grow in the Transvaal, Natal, Rhodesia, Botswana and Malawi.

DESCRIPTION: The height and spread of the species varies considerably. They do not bear showy flowers but are worth growing for the foliage or because of their decorative winged seeds.

CULTURE: These trees are not quick-growing and they do best in regions where winters are not very severe. They can endure long periods with little water.

T. mollis

This species is to be found in Rhodesia growing to a height of 9-10 m (30-40 ft). It is a deciduous tree with large, oval leaves, 12-25 cm long and 5-10 cm broad. The flowers are inconspicuous but the pale-green seeds which follow make a pretty show in summer.

T. prunioides HARDEKOOLBOOM

Grows to 6-8 m (20-26 ft) or more, but may remain much smaller and of shrubby growth. The leaves are broader at the apex than the base and are 2-4 cm long and up to 3 cm broad. The fruits, which appear in summer, hang down in clusters and become prettily tinted from rose to maroon.

T. sericea TRANSVAAL SILVER TREE

This is a small, deciduous tree growing to 8 m (26 ft) although in Malawi, specimens 12 m in height may be found. The leaves are obovate and crowded at the ends of stems. They measure 5-15 cm in length and are covered with silvery hairs, particularly when young, which make them glisten in the sunlight. The claret-coloured winged seeds are carried in showy clusters in summer. This tree, which is to be found in Botswana, the northern Cape, parts of the Transvaal and Rhodesia does well in sandy soils and in dry places.

TREMA ORIENTALIS PIGEON WOOD,
(T. guineensis) HOPHOUT

DISTRIBUTION: This tree is found mostly in the forests of Natal and the Transvaal.

DESCRIPTION: It is an attractive tree related to the white stinkwood and rather like it in appearance. It has a smooth grey bark and pale green foliage with neatly and finely serrated edges to the leaves. Under forest conditions it grows to 18 m (60 ft) but it is unlikely to grow to more than approximately half of this in the average garden. Its flowers are insignificant but the tree is worth growing because of its fine form and pretty foliage. It makes an attractive shade tree.

CULTURE: It needs good soil for its best development although it will make steady growth even under poor conditions. It stands quite severe frost when once it is established.

TRICHILIA ROKA NATAL OR CAPE
(Trichilia emetica) MAHOGANY, CHRISTMAS BELLS, THUNDER TREE, ROOI-ESSENHOUT

DISTRIBUTION: Grows in forests from the eastern Cape up through Natal into the warmer parts of the Transvaal and Rhodesia.

DESCRIPTION: This is a very handsome, evergreen tree which may reach a height of 18 m (60 ft) in forests, but which seldom grows to more than 12 m (40 ft) in the open. It has a large, rounded crown of densely arranged dark green, glossy leaves which make a magnificent canopy. They are divided into leaflets which measure 7-10 cm in length and which are oval with well defined midribs. The small flowers are white and sweetly scented and give the tree its common name of Christmas bells. This is a really splendid tree for

A well-grown White Ironwood *(Vepris undulata)* forms a canopy of shade.

large gardens and parks and for lining streets and avenues. In regions with warm winters and a high rainfall it is a fine tree for avenue or street planting. Some African tribes use an infusion of the bark as a purgative or emetic. Leaves placed in the bed are thought to act as a soporific.

CULTURE: The Natal mahogany does best where winters are mild and where the rainfall is above 500 mm a year. It can however be grown in gardens where moderate frosts are experienced, provided it is protected during the first three to four years. Under good conditions it is fairly fast-growing. One huge tree near Mitchell Park in Durban has a trunk with a circumference of 6 m (20 ft) and the spread of its crown is over 30 m (100 ft).

TRIMERIA ROTUNDIFOLIA WILD MULBERRY, WILDEMOERBEI
(T. grandifolia)

DISTRIBUTION: Grows from Knysna to Natal and East Griqualand from the coast to an elevation of 1200 m.

DESCRIPTION: Its growth is variable according to the conditions. Sometimes it remains shrub size but it may, in warm humid areas, develop into a tree of 12 m (40 ft) or more. It is usually evergreen but occasionally deciduous. The large handsome leaves are a striking feature of the tree. They are carried in a flat plane on the horizontal branches and somewhat resemble the leaves of a mulberry. They are rounded, with a diameter of about 7 cm, and have serrated margins, and are sometimes notched at the apex. The upper surface of the leaves is a rich deep green and the undersides are paler, with the veins prominently displayed. The flowers are inconspicuous and the fruits which follow are not showy, but the tree is worth growing because of its foliage and fine form. It is fairly quick-growing.

CULTURE: Under harsh growing conditions it is unlikely to develop into a large tree, but it stands fairly severe frost. Plant it in good soil and water it regularly and well to promote speedy growth.

VEPRIS UNDULATA WHITE IRONWOOD, WITYSTERHOUT
(V. lanceolata)

DISTRIBUTION: Is found in forests from the south-western Cape north into Natal and the Transvaal.

DESCRIPTION: In forests this tree may grow to 20 m (65 ft) but in the open it tends to be spreading and does not grow more than about 10 m (30-35 ft) with a spread across 15 m. The trunk is smooth and grey and the leaves are divided into three leaflets 6-8 cm long with rather wavy edges. They are glossy on the upper surface and smell of orange peel when bruised. The flowers are inconspicuous. This is a good tree to provide shade in the large garden or to create shade and shelter for stock on farms.

CULTURE: It does best in areas which have mild to moderate winters and where the rainfall is not less than 600 mm a year. It is not suitable for dry gardens or regions where frosts are severe.

The Keurboom *(Virgilia divaricata)* is a very quick-growing tree with ornamental foliage and pretty, scented flowers.

VIRGILIA DIVARICATA KEURBOOM

DISTRIBUTION: Can be found on the edges of forests from the Cape Peninsula to Port Elizabeth.

DESCRIPTION: Keurboom is a very decorative and exceptionally fast-growing tree. In spring it bears showy clusters of sweetly scented flowers which are of a very delightful shade of pinky-mauve to deep mauve. The tree is however, attractive throughout the year as it is an evergreen with graceful foliage. The leaves are divided into small leaflets only 2-3 cm long, which are dark green in colour. The keurboom grows to about 8 m (25 ft) in height with a spread of almost as much, and it is an excellent tree for gardens large and small, and also for avenue planting. It is not a long-lived tree, probably because it grows so rapidly, and it may be necessary to renew the trees every fifteen years. It is also apt to have its branches broken by strong wind or to be blown over. However its general attractiveness and the fact that it grows so quickly outweigh any demerits it may have. *Virgilia oroboides* is so similar to the species described above that some authorities consider that it should not have a separate species name. It flowers in summer, and although its clusters of flowers are not as splendid, nor its foliage quite as lustrous, it is nevertheless a fine tree to grow as it produces its flowers at a different season.

CULTURE: Keurbooms grow best in regions where winters are mild to moderate and where the rainfall is good, or where they can be watered regularly, particularly when young. Once established, however, they tolerate fairly long periods of drought and fairly sharp frosts as well. This is the quickest-growing of all our indigenous trees and, when planted close together, it will make a dense screen, but they look best when planted where they can spread into round-headed trees. When planted in gardens where frosts are fairly severe the young trees should be protected during winter for the first two or three years. They are not suitable trees for gardens where frosts are really severe and the air dry throughout the year.

ZIZIPHUS MUCRONATA BUFFALO-THORN
BLINKBLAAR-WAG-'N-BIETJIE

DISTRIBUTION: Is to be found in all the provinces of South Africa, in Botswana, South West Africa and further north in Rhodesia.

DESCRIPTION: It is variable in growth—sometimes being no larger than a shrub and sometimes a tree to 9 m (30 ft), with a rounded spreading crown. It is deciduous or semi-deciduous with broadly oval leaves ending in a point. These vary in size according to habitat but they always have their characteristic veining—three main ones curving up from the base of each leaf. They may measure 3-6 cm in length and 2-4 cm across, and they have neatly serrated margins and a pleasing glossy appearance. This shining quality of the leaves and the small strong straight and hooked thorns on the stems accounts for the common name of the tree. Once caught by the thorns it is difficult to extricate oneself. The small greenish-yellow flowers which appear in spring are not conspicuous. They are followed by fruits which ripen in autumn to a reddish-brown colour. These were relished by native tribes as a form of food, and during the Boer War commandos on the move roasted the dried fruits to use as a substitute for coffee. In early days in the Transvaal the settlers cut these trees in large numbers to make kraals or pallisades to protect their stock from lions. The leaves are eaten by buck and by stock and various parts of the tree are used by African people medicinally. An infusion of the roots is still used to cure stomach ailments and for treating boils. Some tribes also attribute magical qualities to this tree, believing that anyone standing under the tree will not be struck by lightning.

CULTURE: This tree should be more widely grown for it is decorative when well grown, and it survives difficult growing conditions—such as poor soil, drought, cold and heat. Under optimum conditions it is fairly fast in growth.

Karee *(Rhus lancea)* is a good tree to provide shade and shelter in dry regions which experience sharp frost.

(Page 207.)

The Buffalo Thorn or Blinkblaar-wag-'n-Bietjie *(Ziziphus mucronata)* is another tree which tolerates difficult growing conditions.

Index to Common Names

A Aardroos, 157
Abiekwas, 211
African
  Beech, 193
  Dog Rose, 138
  Flame Tree, 211
  Wattle, 204
Alberta, 182
Amatungulu, 74
Anaboom, 178
Anatree, 178
Apiesdoring, 178, 179
Ash
  Cape, 190
  Rhodesian, 185
Assegaaihout, 188
Assegaai Wood, 111
Assegaaiwood, 188
Aulax, 70

B Baakhout, 195
Bakhout, 200
Baobab, 181
Barleria, 70
Bastard Willow, 207
Basterkameel, 179
Bergaster, 114
Bergroos, 157
Bird Flower, 77
Bitter Almond, 205
Black
  Bark, 190
  Ironwood, 203
Black-eyed Susan, 168
Blinkblaarboom, 206
Blinkblaar-wag-'n-bietjie, 214
Blombos, 133
Bloodwood, 206
Bloukappie, 143
Blue Boys, 162
Blushing Bride, 164, 165
Bobbejaanstou, 163
Boekenhout, 193, 206
Boerboon, 207
  Forest, 208
  Karoo, 208
  Weeping, 208
Boetabessie, 75
Bostaaibos, 206
Bottlebrush
  Mountain, 195
  Natal, 195
  Red-and-yellow, 135
Bread Tree, 80
Bredasdorp Protea, 158
Broodboom, 80
Brunia, 72, 73
Buchu
  Oval-leaf, 67
Buffalo Thorn, 214
Burning Bush, 76
Bushman's Poison Bush, 65
Bush Clematis, 75
Bush Tea, 77
Bush-Willow, 187
Buttonwood, 111

C Cabbage Tree, 188
Calpurnia, 74
Camdeboo Stinkwood, 186
Camel Thorn, 179
Camphor Wood, 211
Canary Creeper, 164
Cancer Bush, 167
Cape
  Ash, 190
  Beech, 206
  Chestnut, 186
  Fig, 193
  Holly, 197
  Honeysuckle, 168
  Laburnum, 77
  Leadwort, 141
  Mahogany, 212
  Trumpet Flower, 168
  Willow, 207
Carnival Bush, 136
Cassia, long-tail, 186
Catherine Wheel, 122
China Flower, 66, 67
Chinese Lantern, 136
Christmas
  Bells, 212
  Bush, 139
Cinderella, 164
Clematis, Bush, 75
Coleonema, 76
Confetti Bush, 76
Coulter Bush, 69
Cream of Tartar Tree, 181
Crossberry, 111
Cucumber Tree, 198
Curry Bush, 112
Cycad
  Modjadji, 81
  Mujaji, 81
  Natal, 81
  Prickly, 81

D Dahlia, mountain, 131
Dais, 189
Daisy Bush, 108, 109
Desert
  Broom, 74
  Rose, 112
Dikbas, 190
Dikbos, 112
Dingaan's Apricot, 79
Dog Plum, 190
Dombeya, 79, 190
  Heart-leaf, 79
  Pink and White, 79
Doring
  Apies, 178
  Enkel, 180
  Gousblom, 71
  Kameel, 179
  Kat, 179
  Knoppies, 180
  Ou, 180
  Platkroonsoet, 183
  Soet, 179
  Vaalkameel, 179
Duinebessie, 135
Dwarf
  China Flower, 66
  Green Protea, 148
  Kaffirboom, 107

E Eight-day-healing Bush, 131
Enkeldoring, 180
Erica, 82
  Beaked, 96
  Daphne, 89
  Fairy, 86
  Red Hairy, 88
  Woolly, 98
  see also under Heath
Erythrina, 191
Essenhout, 190
Euryops, 109
  Grey-leafed, 109

F Feather Climber, 65
Featherhead, 139
Fevertree, 181
Fig
  Cape, 193
  Natal, 193
  Red-leafed, 193
  Wild, 193
Fire Wheel Pincushion, 127
Flame Creeper, 77
Flat-crown, 183

Fluitjies, 114
Forest
Boerboon, 208
Primrose, 112
Waterwood, 211
Freylinia, 193
Fuchsia
Tree, 208

G Gansies Keur, 167
Gardenia, 194
East London, 111
Transvaal, 194
Wild, 194
Geelbos, 120
Geelgranaat, 16
Geelkoppie, 131
Gifboom, 65
Gold Tips, 121
Ground Rose, 157
Grysappel, 203
Gwarrie, 192

H Haak-en-steek, 180
Hardekoolboom, 212
Hard Pear, 203
Heath, 82, 105
Albertinia, 86
Berry, 84
Blandford's, 86
Bottle, 84
Bridal, 86
Cup-and-saucer, 93
Double Pink, 102
Dwarf, 97
Elim, 102
Flounced, 87
Four Sisters, 90
Franschhoek, 103
Funnel, 94
Gansbaai, 94
Globe, 96
Grahamstown, 89
Green, 102, 105
Green-edged, 89
Hedge, 87
Houhoek, 97
Knysna, 90
Lantern, 86
Large Orange, 93
Long-leafed, 96
Malay, 102
Mealie, 98
Mountain, 103
Orange, 86
Petticoat, 93
Pillans, 99
Pink Shower, 102
Port Elizabeth, 89
Prince of Wales, 99
Red Signal, 97
Riversdale, 86
Royal, 102
Scented Bell, 89

Heath
Sissie, 84
Small Flounced, 94
Small Tassel, 89
Star-faced, 102
Sticky-leafed, 93
Sticky Rose, 89
Swellendam, 105
Tassel, 99
Tigerhoek, 90
Velvet Bell, 99
Walker's, 105
Water, 89
Wax, 84, 103
Wide-mouthed, 105
Wood's, 105
Yellow Bell, 87

Hermannia, 111
Heuningtee, 77
Hiccup-nut, 76
Honey Flower, 131
Dwarf, 133
Large, 133
Honeysuckle
Cape, 168
Honey Tea, 77
Hophout, 212
Hottentot's Bean, 207, 208
Huilboom, 204
Huilboerboon, 208

I Iboza, 112
Ironwood
Black, 203
Kaffir, 201
White, 213

J Jasmine
East London, 113
Natal, 114
Jasmine Tree, 197

K Kaboom, 179
Kaffir
Ironwood, 201
Thorn, 179
Wag-'n bietjie, 179
Kaffirboom, 191
Broad-leaf, 192
Dwarf, 107
Kaffirbread Tree, 82
Kaffirplum, 196
Kaffirthorn, 179
Kamassie, 194
Kameeldoring, 179
Kankerbos, 167
Kannabas, 189
Kapokbossie, 105
Kapokkie, 99
Karee, 207
Karoo Boerboon, 208

Karoo
Gold, 163
Tree, 207
Kasuur, 204
Katdoring, 179
Katjiepiering, 194
Wilde, 194
Kei Apple, 79
Kerriebos, 112
Keurboom, 214
Keurtjie, 141, 143
Kiaat
Round-leaf, 206
Kiepersol, 188
Kierieklapper, 76
Klaas Louw Bos, 69
Klapperbos, 136
Klimop, 75
Knobkerrie Bush, 120
Knob Thorn, 180
Knobwood, 192
Knopdoring, 192
Knoppiesdoring, 180
Knysna Boxwood, 194
Kolkol, 72
Kommetjieteewater, 67
Koulterbos, 69
Kreupelboom, 137
Kreupelhout, 125
Kruidjie-roer-my-nie, 131
Kruisbessie, 111

L Lachnaea, 114
Lebeckia, 114
Dwarf, 115
Lekkerbreek, 202
Lemonade Tree, 181
Leucadendron, 115-120
Leucospermum, 121-129
Live Long, 200
Long-tail Cassia, 186
Lowveld Chestnut, 211
Lucky Bean Tree, 191
Luisies, 121

M Mackaya, 131
Mahogany Bean, 182
Mauve Chinese Hat Plant, 112
Melkbos, 67
Metalasia, 133
Mauve, 135
Milkwood Tree, 201
Mimetes, 135
Silver-leaf, 135
Mimosa, 179
Mobala Plum, 203
Moepel, 202
Monkey Thorn, 179
Monkey-rope, 163
Mountain
Ash, 190
Bottlebrush, 195
Dahlia, 131

Mountain
  Daisy, 108
  Rose, 157
Msasa, 184
Munjerenje, 183
Mustard Bush, 112

N Natal
  Bottlebrush, 195
  Camelthorn, 180
  Fig, 193
  Mahogany, 212
  Plum, 74
Nchongo, 206
Notsung, 196
Num-num, 74
Nyala, 206

O Oak, sea, 208
Ochna, 136, 202
Old Man's Beard, 75
Olienhout, 203
Olive
  Sand, 190
  Wild, 203
Olyf
  Sand, 190
  Wit, 203
Oncoba, 138
Oudoring, 180
Oval Leaf Buchu, 67

P Paperbark Thorn, 180
Paranomus
  Green, 138
Perdebos, 138
Pigeon Wood, 212
Pincushion, 121-127
  Creeping, 127
  Fire Wheel, 127
  Muir's, 127
  Narrow-leafed, 125
  Rainbow, 125
  Rocket, 127
  Tufted, 127
  Upright, 127
  White, 122
Pink
  Coleonema, 76
  Plume, 167
Pistol Bush, 80
Platkroon, 183
Ploegbreker, 107
Pluimbossie, 75
Plumbago, 141
Polygala, 143
Pompon Tree, 189
Porcupine salvia, 162
Port St. John's Climber, 143
Prickly
  Cardinal, 107
  Sunflower, 71
Pride of
  De Kaap, 71
  Franschhoek, 167
Protea, 144-162
  Baby, 155
  Bearded, 151
  Black-bearded, 156
  Bot River, 152
  Bredasdorp, 158
  Brown-bearded, 161
  Cedarberg, 151
  Christmas, 151
  Drakensburg, 161
  Dwarf Green, 148
  Fringed, 155
  Giant, 152
  Giant Woolly, 151
  Gleaming, 161
  Green, 156
  Hanging, 158
  Highveld, 151
  King, 152
  Ladismith, 151
  Long-bud, 156
  Long-leafed, 156
  Oleander-leafed, 157
  Peach, 155
  Pine-leafed, 158
  Queen, 151
  Ray-flowered, 155
  Small, green, 161
  Snow, 152
  Sprawling, 150
  Stokoe's, 162
  Susan's, 162
  Swartberg, 162
Purple Broom, 143, 144

Q Quinine Tree, 206

R Rafinia, 163
Red
  Alder, 187
  -and-yellow Bottlebrush, 135
  Currant, 206
  -leafed Fig, 193
  Mahogany, 198
  Milkwood, 202, 205
  Stinkwood, 205
Resin Bush, 108, 111
Rhodesian
  Ash, 185
  Black Wattle, 204
  Chestnut, 183
  Ebony, 190
  Flame Tree, 211
  Mahogany, 182
  Teak, 183
River Bells, 139
Rooibessie, 203
Rooi-els, 187
Rooi-essenhout, 212
Rooi-opslag, 112
Rooistompie, 135
Rooiwildevy, 193
Roundleaf Kiaat, 206
Russet Bush Willow, 76

S Saffraan, 186
Saffron, 186
Sage Wood, 184
Saliehout, 184
Salvia
  Porcupine, 162
Sand Olive, 190
Sandolyf, 190
Sausage Tree, 198
Scented
  Bell Heath, 89
  Cups, 163
Sea Oak, 208
September Bells, 163
Septemberbossie, 144
Seringa
  White, 198
  Wild, 185
Serruria, 164-167
  Grey, 167
  Silky, 164
Shepherd's Delight, 66
Shock-headed Peter, 75
Silver
  Pea, 144
  Sweetpea, 143
  Tree, 200
Skaamroos, 157
Slangbos, 167
Small Flounced Heath, 94
Snow Protea, 152
Soetdoring, 179
Soldaat, 135
South African Gorse, 68
Speldekussing, 125
Spiderbush, 167
Spinnekopbos, 167
Spur Flower, 140
  Pink, 140
  Purple, 140
Stinkwood
  Camdeboo, 186
  Red, 205
  White, 186
Stinkhout
  Wit, 186
Stock Rose, 209
Stompie, 72, 73
Sugarbush, 161
Suikerbos, 161
Sulphur Bark, 181
Swartberg Protea, 162
Swartbas, 190
Swartstoom, 74
Sweet Thorn, 179
Sweetpea Bush, 141
  Silver, 143

T
Tamboekie Thorn, 107
Terblans, 192
Thorn, 178
Camel-, 179
Kaffir, 179
Knob-, 180
Monkey, 179
Natal Camel-, 180
Paperbark, 180
Sweet-, 179
Tamboekie, 107
Umbrella, 180
White, 178
Thunder Tree, 212
Tierhout, 201
Tortoise Berry, 135
Transvaal
Ebony, 190
Gardenia, 194
Red Milkwood, 202
Silver Tree, 212
Teak, 206
Traveller's Joy, 75
Tree
Fuchsia, 196, 208
Wistaria, 183
Trots van Franschhoek, 167

U
Umbrella
Thorn, 180
Tree, 189
Umkwenkwe, 204
Umzimbeet, 201

V
Vaalbos, 184, 211
Vaalkameeldoring, 179
Vaderlandswilg, 187
Van Wykshout, 183
Vlier, 202
Volstruiskameel, 179

W
Waboom, 150
Waterboom, 197
Waterbos, 89
Water Pear, 211
Wattle
African, 204
Rhodesian Black, 204
Weeping Boerboon, 208
White
Elder, 202
Ironwood, 213
Milkwood, 208
Pear, 183
Seringa, 198
Stinkwood, 186
White Thorn, 178
Wild
Almond, 184
Aster, 68
Broom, 114
Chestnut, 186
Cotton, 67
Date, 82
Fig, 193
Gardenia, 194
Grape, 163
Lasiandra, 78
Mulberry, 213
Olive, 203
Peach, 198
Pear, 190
Pepper Tree, 201
Wild
Plum, 190
Pomegranate, 73
Rosemary, 105
Sage, 211
Seringa, 185
Tamarisk, 211
Willow, 207
Wilde-amandel, 184
Wildebesembos, 114
Wildegranaat, 73
Wildekapok, 67
Wildemoerbei, 213
Wildeperske, 198
Willow
Bastard, 207
Cape, 207
Wild, 207
Witolyf, 196
Witpeer, 183
Witsering, 198
Witstinkhout, 186
Witysterhout, 213
Wonderboom, 193
Wood's Heath, 105

Y
Yellowwood
Common, 205
Henkel's 205
Outeniqua, 205
Real, 205
Upright, 205
Ysterhout, 190, 203
Wit, 213

Z
Zambezi Redwood, 183
Zimbabwe Creeper, 143

Index to Botanical Names

A Acacia albida, 178
caffra, 179
galpinii, 179
giraffae, 179
haematoxylon, 179
karroo, 179
nigrescens, 180
robusta, 180
sieberiana var. woodii, 180
tortilis subsp. heteracantha, 180
xanthophloea, 181
Acokanthera oppositifolia, 65
venenata, 65
Acridocarpus natalitius, 65
Adansonia digitata, 181
Adenandra cuspidata, 66
fragrans, 66
serphyllacea, 66
umbellata, 66
uniflora, 67
Adhatoda duvernoia, 80
Afzelia cuanzensis, 182
Agathosma crenulata, 67
Alberta magna, 182
Albizia adianthifolia 183
gummifera, 183
harveyi, 183
Apodytes dimidiata, 183
Asclepias fruticosa, 67
physocarpa, 67
Aspalathus capensis, 65
macrantha, 68
sarcodes, 68
spinosa, 68
Aster filifolius, 68
Athanasia acerosa, 69
crithmifolia, 69
parviflora, 69
Aulax cneorifolia, 70
pinifolia, 70

B Baikiaea plurijuga, 183
Barleria obtusa, 70
Barosma crenulata, 67
Bauhinia galpinii, 71
Berkheya barbata, 71
ilicifolia, 71
Berzelia lanuginosa, 72
Bolusanthus speciosus, 183
Brabejum stellatifolium, 184
Brachylaena discolor, 184
Brachystegia spiciformis, 184
Brunia laevis, 73
nodiflora, 73
paleacea, 136
stokoei, 73
Buddleia salviifolia, 184
Burchellia bubalina, 73
Burkea africana, 185

C Cadaba aphylla, 74
Calodendrum capense, 186
Calpurnia floribunda, 74
intrusa, 74
villosa, 74
Carissa bispinosa, 75
grandiflora, 75
macrocarpa, 75
Cassia abbreviata subsp.
bearana, 186
Cassine capensis, 186
crocea, 186
peragua, 186
Celtis africana, 186
kraussiana, 186
Chrysanthemoides monilifera, 75
Clematis brachiata, 75
Clematopsis scabiosifolia, 75
stanleyi, 75
Coleonema album, 76
pulchrum, 76
Combretum bracteosum, 76
erythrophyllum, 187
hereroense, 76
kraussii 187
microphyllum, 77
transvaalense, 76
zeyheri, 187
Crotalaria agatiflora, 77
capensis, 77
mucronata, 77
purpurea, 77
striata, 77
Cunonia capensis, 187
Curtisia dentata, 188
Cussonia paniculata, 188
spicata, 189
Cyclopia genistoides, 77

D Dais cotinifolia, 189
Diospyros mespiliformis, 190
whytei, 190
Diplopappus filifolius, 68
Dissotis canescens, 78
incana, 78
princeps, 78
Dodonaea thunbergiana, 190
viscosa, 190
Dombeya burgessiae, 79
dregeana, 79
rotundifolia, 190
tiliacea, 79
Dovyalis caffra, 79
Duvernoia adhatodioides, 80

E Ekebergia capensis, 190
meyeri, 191
Encephalartos altensteinii, 81
ferox, 81
horridus, 81
natalensis, 81
transvenosus, 81
villosus, 82
Erica ampullacea, 84
ardens, 84
aristata, 84
baccans, 84
bauera, 86
bergiana, 86
blandfordia, 86
blenna, 86
blenna var. grandiflora, 87
bodkinii, 87
borboniaefolia, 87
bowieana, 86
caffra, 87
campanularis, 87
cerinthoides, 88
chamissonis, 89
chloroloma, 89
coccinea, 89
coccinea var. melastoma, 89
corifolia, 89
curviflora, 89
curvirostris, 89
daphniflora, 89
decora, 89
deliciosa, 89
densifolia, 90
eugenea, 90
fascicularis, 90
fastigiata, 90
formosa, 93
glandulosa, 93
glauca, 93
glauca var. elegans, 93
grandiflora, 93
grandiflora var. exsurgens, 93
grisbrookii, 93
haematosiphon, 93
hibbertia, 94
holosericea, 94
imbricata, 94
inflata, 94
infundibuliformis, 94
irregularis, 94
junonia, 94
lanipes, 96
lanuginosa, 96
lateralis, 96
longifolia, 96
lutea, 96
macowanii, 96
mammosa, 97
massoni, 97
nana, 97

Erica
oatesii, 97
ovina, 98
ovina var. *purpurea, 98*
parilis, 98
passerina, 98
patersonia, 98
perspicua, 99
petiveri, 89
peziza, 99
pillansii, 99
pinea, 99
plukeneti, 99
porteri, 102
quadrangularis, 102
regia, 102
regia var. *variegata, 102*
sessiliflora, 102
shannonea, 102
sparrmanni, 102
taxifolia, 102
thunbergii, 102
tumida, 103
ventricosa, 103
versicolor, 105
vestita, 105
viridiflora, 105
walkeria, 105
woodii, 105
Eriocephalus africanus, 105
punctulatus, 106
Eroeda imbricata, 106
Erythrina abyssinica, 192
acanthocarpa, 107
caffra, 191
humeana, 107
latissima, 192
lysistemon, 192
zeyheri, 107
Euclea crispa, 192
Eugenia gerrardi, 211
Euryops abrotanifolius, 109
acraeus, 109
athanasiae, 111
chrysanthemoides, 109
linifolia, 109
pectinatus, 109
speciosissimus 111
tenuissimus, 111
thunbergii, 111
unnamed hybrid, 111
virgineus, 111

F
Fagara daryi, 192
Faurea galpinii, 193
macnaughtonii, 192
saligna, 193
Ficus capensis, 193
ingens, 193
natalensis, 193
pretoriae, 193
Freylinia lanceolata, 193

G
Gamolepis chrysanthemoides, 109
Gardenia amoena, 111
capensis, 164
gerrardiana, 111
globosa, 164
spatulifolia, 194
thunbergia, 194
Gonioma kamassi, 194
Grewia occidentalis, 111
Greyia flanaganii, 196
radlkoferi, 196
sutherlandii, 195

H
Halleria lucida, 196
Harpephyllum caffrum, 196
Hermannia althaeifolia, 112
stricta, 112
vesicaria, 112
Holarrhena febrifuga, 197
Holmskioldia speciosa, 112
tettensis, 112
Hypericum leucoptycodes, 112
revolutum, 112

I
Iboza riparia, 112
Ilex capensis, 197
mitis, 197
Indigofera cuneifolia, 113
cylindrica, 113
cytisoides, 113
filifolia, 113
frutescens, 113

J
Jasminum angulare, 113
breviflorum, 114
gerrardii, 114
multipartitum, 114

K
Khaya nyasica, 198
Kigelia africana, 198
pinnata, 198
Kiggelaria africana, 198
Kirkia acuminata, 198
wilmsii, 200

L
Lachnaea buxifolia, 114
densiflora, 114
diosmoides, 114
Lachnopylis floribunda, 202
Lannea discolor, 200
Lebeckia cytisoides, 114
simsiana, 115
Leucadendron adscendens, 120
aemulum, 116
album, 116
argenteum, 200
aurantiacum, 116
chamelaea, 116
comosum, 116
conicum, 116
coniferum, 116
daphnoides, 116
decorum, 118
decurrens, 116
discolor, 116
Leucadendron
elimense, 116
eucalyptifolium, 118
floridum, 118
galpinii, 118
gandogeri, 118
glabrum, 121
grandiflorum, 121
guthrieae, 118
laureolum, 118
loranthifolium, 118
macowanii, 118
microcephalum, 119
modestum, 119
muirii, 120
nervosum, 120
platyspermum, 120
plumosum, 120
pubescens, 120
rubrum, 120
sabulosum, 116
salicifolium, 120
salignum, 120
seriocephalum, 120
sessile, 121
spissifolium, 121
stokoei, 119
strictum, 120
tinctum, 121
uliginosum, 121
venosum, 121
xanthoconus, 121
Leucospermum album, 122
attenuatum, 125
bolusii, 122
buxifolium, 127
calligerum, 122
candicans, 127
catherinae, 122
conocarpodendron 125
conocarpum, 125
cordifolium, 125
crinitum, 127
cuneiforme, 125
glabrum, 125
grandiflorum, 125
gueinzii, 125
incisum, 127
lineare, 125
muirii, 127
mundii, 127
nutans, 125
oleifolium, 127
prostratum, 127
puberum 122
reflexum, 127
rodolentum, 127
tottum, 127
truncatulum, 127
vestitum, 127
Liparia sphaerica, 131
splendens, 131
Lobostemon fruticosus, 131
trichotomus, 131

Lobostemon
trigonus, 131
Loxostylis alata, 201

M
Mackaya bella, 131
Melianthus dregeana, 131
insignis, 133
major, 133
minor, 133
Metalasia aurea, 133
muricata, 133
rhoderoides, 133
seriphiifolia, 135
Millettia caffra, 201
grandis, 201
sutherlandii, 201
Mimetes argenteus, 135
cucullatus, 135
hirtus, 135
lyrigera, 135
Mimusops caffra, 202
obovata, 202
zeyheri, 202
Mundia spinosa, 135

N
Nebelia paleacea, 136
Nuxia floribunda, 202
Nymania capensis, 136

O
Ochna arborea, 202
atropurpurea, 136
pulchra, 202
Oldenburgia arbuscula, 137
Olea africana, 203
capensis subsp. macrocarpa, 203
Olinia cymosa, 203
emarginata, 203
radiata, 203
Oncoba kraussiana, 138
spinosa, 138

P
Paranomus reflexus, 138
spicatus, 139
Parinari curatellifolia, 203
Pavetta lanceolata, 139
obovata, 139
revoluta, 139
Peltophorum africanum, 204
Phygelius capensis, 139
equalis, 139
Phylica pubescens, 139
Pittosporum viridiflorum, 204
Plectranthus behrii, 140
ecklonii, 140
Plumbago auriculata, 141
capensis, 141
Podalyria burchellii, 141
calyptrata, 143
sericea, 143
Podocarpus falcatus, 205
henkelii, 205
latifolius, 205
Podranea brycei, 143
ricasoliana, 143
Polygala myrtifolia, 144
virgata, 144
Priestleya hirsuta, 144
tomentosa, 144
villosa, 144
Protea acaulis, 148
acerosa, 150
amplexicaulis, 150
arborea, 150
aristata, 151
aspera, 151
barbigera, 151
caffra, 151
cedromontana, 151
compacta, 152
convexa, 152
cryophila, 152
cynaroides, 152
decurrens, 152
effusa, 152
eximia, 155
grandiceps, 155
grandiflora, 150
harmeri, 155
incompta, 156
lacticolor, 155
lanceolata, 155
latifolia, 155
laurifolia, 155
lepidocarpodendron, 156
longiflora, 156
longifolia, 156
lorifolia, 156
macrocephala, 156
macrophylla, 156
marginata, 155
marlothii, 152
mellifera, 161
minor, 157
mundii, 157
nana, 157
neriifolia, 157
obtusifolia, 158
pendula, 158
pityphylla, 158
pulchra, 161
punctata, 155
repens, 161
rosacea, 157
rouppelliae, 161
rupicola, 161
scolymocephala, 161
speciosa, 161
speciosa var. angustata, 162
stokoei, 162
subpulchella, 161
sulphurea, 162
susannae, 162
venusta, 162
Prunus africanum, 205
Pseudocadia zambesiaca, 206
Pterocarpus angolensis, 206
rotundifolius, 206
Pycnostachys urticifolia, 162
Pygeum africanum, 205

R
Rafnia ovata, 163
thunbergii, 163
Rapanea melanophloeos, 206
Rauvolfia caffra, 206
Rhigozum brevispinosum, 163
obovatum, 163
Rhoicissus capensis, 163
tomentosa, 163
Rhus chirindensis, 206
lancea, 207
leptodictya, 207
viminalis, 207
Rothmannia capensis, 164
globosa, 164
Royena lucida, 190

S
Salix capensis, 207
hirsuta, 207
mucronata, 207
woodii, 207
wilmsii, 207
Schotia afra, 208
brachypetala, 208
latifolia, 208
Senecio tamoides, 164
Serruria adscendens, 164
aemula, 164
artemisiaefolia, 167
barbigera, 164
burmanii, 167
florida, 167
pedunculata, 167
Sideroxylon inerme, 208
Sparrmannia africana, 209
Spathodea campanulata, 211
Sterculia murex, 211
Stoebe plumosa, 167
Sutherlandia frutescens, 167
Syncolostemon densiflorus, 167
Syzgium cordatum, 211
gerrardii, 211

T
Tamarix austro-africana, 211
Tarchonanthus camphoratus, 211
Tecomaria capensis, 168
capensis var. lutea, 168
Terminalia mollis, 212
prunioides, 212
sericea, 212
Thunbergia alata, 168
Trema guineensis, 212
orientalis, 212
Trichilia emetica, 212
roka, 212
Trimeria grandifolia, 213
rotundifolia, 213

V
Vepris lanceolata, 213
undulata, 213
Virgilia divaricata, 214
oroboides, 214

Z
Ziziphus mucronata, 214